Springer Theses

Recognizing Outstanding Ph.D. Research

For further volumes:
http://www.springer.com/series/8790

Aims and Scope

The series "Springer Theses" brings together a selection of the very best Ph.D. theses from around the world and across the physical sciences. Nominated and endorsed by two recognized specialists, each published volume has been selected for its scientific excellence and the high impact of its contents for the pertinent field of research. For greater accessibility to non-specialists, the published versions include an extended introduction, as well as a foreword by the student's supervisor explaining the special relevance of the work for the field. As a whole, the series will provide a valuable resource both for newcomers to the research fields described, and for other scientists seeking detailed background information on special questions. Finally, it provides an accredited documentation of the valuable contributions made by today's younger generation of scientists.

Theses are accepted into the series by invited nominated only and must fulfill all of the following criteria

- They must be written in good English.
- The topic of should fall within the confines of Chemistry, Physics and related interdisciplinary fields such as Materials, Nanoscience, Chemical Engineering, Complex Systems and Biophysics.
- The work reported in the thesis must represent a significant scientific advance.
- If the thesis includes previously published material, permission to reproduce this must be gained from the respective copyright holder.
- They must have been examined and passed during the 12 months prior to nomination.
- Each thesis should include a foreword by the supervisor outlining the significance of its content.
- The theses should have a clearly defined structure including and introduction accessible to scientists not expert in that particular field.

Emanuel Schneck

Generic and Specific Roles of Saccharides at Cell and Bacteria Surfaces

Revealed by Specular and Off-Specular X-Ray and Neutron Scattering

Doctoral Thesis accepted by the University of Heidelberg, Germany

 Springer

Author
Dr. Emanuel Schneck
Physical Chemistry of Biosystems
  and BIOQUANT
University of Heidelberg
lm Neuenheimer Feld 253
69120 Heidelberg
Germany
e-mail: emanuel.schneck@ph.tum.de

Supervisor
Prof. Motomu Tanaka
Physical Chemistry of Biosystems
  and BIOQUANT
University of Heidelberg
lm Neuenheimer Feld 253
69120 Heidelberg
Germany
e-mail: tanaka@uni-heidelberg.de

ISSN 2190-5053

e-ISSN 2190-5061

ISBN 978-3-642-15449-2

e-ISBN 978-3-642-15450-8

DOI 10.1007/978-3-642-15450-8

Springer Heidelberg Dordrecht London New York

Cover Design: eStudio Calamar, Berlin/Figueres

Printed on acid-free paper

Springer is part of Springer Science+Business Media (www.springer.com)

In any given case it would doubtless prove possible to carry out the analysis numerically if not algebraically, but with such matters we are not here concerned.

David C. Grahame, 1953

Parts of this Thesis have been Published in the Following Journal Articles:

E. Schneck, F. Rehfeldt, R.G. Oliveira, C. Gege, B. Demé, M. Tanaka, *Modulation of intermembrane interaction and bending rigidity of biomembrane models via carbohydrates investigated by specular and off-specular neutron scattering.* Phys. Rev. E, **78**, 061924 (2008)

E. Schneck, E. Papp-Szabo, B.E. Quinn, O.V. Konovalov, T.J. Beveridge, D.A. Pink, M. Tanaka, *Calcium ions induce collapse of charged O-side chains of lipopolysaccharides from Pseudomonas aeruginosa.* J. R. Soc. Interface, **6**, S671 (2009)

E. Schneck, R.G. Oliveira, F. Rehfeldt, B. Demé, K. Brandenburg, U. Seydel, M. Tanaka, *Mechanical Properties of Interacting Lipopolysaccharide Membranes from Bacteria Mutants Studied by Specular and Off-Specular Neutron Scattering.* Phys. Rev. E **80**, 041929 (2009)

E. Schneck, T. Schubert, B.E. Quinn, O.V. Konovalov, D.A. Pink, M. Tanaka, *Quantitative Determination of Ion Concentration Profiles at Bacteria Surfaces Revealed by Grazing-Incidence X-Ray Fluorescence.* Proc. Natl. Acad. Sci. **107**, 9147 (2010)

R.G. Oliveira, E. Schneck, K. Brandenburg, U. Seydel, B.E. Quinn, O.V. Konovalov, T. Gill, D.A. Pink, M. Tanaka, *Physical mechanism of bacterial survival revealed by combined grazing-incidence X-ray scattering and Monte Carlo simulation.* Comptes Rendus Chimie **12**, 209–217 (2009)

R.G. Oliveira, E. Schneck, B.E. Quinn, O.V. Konovalov, K. Brandenburg, T. Gutsmann, T. Gill, C.B. Hanna, D.A. Pink, M. Tanaka, *Crucial roles of charged saccharide moieties in survival of Gram-negative bacteria revealed by combination of grazing Incidence x-ray structural characterizations and Monte Carlo simulations.* Phys. Rev. E **81**, 041901 (2010)

T. Schubert, P. Seitz, E. Schneck, M. Nakamura, M. Shibakami, S.S. Funari, O.V. Konovalov, M. Tanaka, *Structure of synthetic transmembrane lipid membranes at the solid/liquid interface studied by specular x-ray reflectivity.* J. Phys. Chem. B **112**, 10041 (2008)

M. Tanaka, M. Tutus, S. Kaufmann, F.F. Rossetti, E. Schneck, I. Weiss, *Native supported membranes on planar polymer supports and micro-particle supports.* J. Struct. Biol. **168** (1) (2009) review article

M. Tutus, F.F. Rossetti, E. Schneck, G. Fragneto, F. Förster, R. Richter, T. Nawroth, M. Tanaka, *Orientation-Selective Incorporation of Transmembrane F0F1 ATP Synthase Complex from Micrococcus luteus in Polymer-Supported Membranes*. Macromolecular Bioscience **8**, 1034 (2008)

F.F. Rossetti, P. Panajiotou, F. Rehfeldt, E. Schneck, M. Dommach, S.S. Funari, A. Timmann, P. Müller-Buschbaum, M. Tanaka, *Structure of regenerated cellulose films revealed by grazing incidence small-angle X-ray scattering*. Biointerphases **3**(4), 117–127 (2008)

Supervisor's Foreword

Membranes are key components of all biological systems, defining the boundary between cytoplasmic (interior) and extracellular (exterior) spaces. In particular, membrane proteins and carbohydrates attached to the membrane surface (glycocalix) are important regulators in intercellular communication and transport across the membrane, so that membranes act as filters and platforms for a variety of biochemical processes. In nature, cell–cell, and cell–tissue interactions are mediated via glycocalix and extracellular matrix, which are mainly composed of oligo- and polysaccharides. Also, saccharides can act as specific ligands for proteins and complementary saccharides. Due to this complexity, it is essential to design well-defined model systems with a reduced number of components in order to quantitatively understand the specific and generic roles of saccharides in modulating interactions at biological interfaces. The main thrust of the Ph.D. thesis of Emanuel Schneck was to quantitatively investigate the generic and specific roles of saccharide chains coupled to the membrane surface in the modulation of interactions between cells/bacteria and their environments. To probe the fine structures perpendicular and parallel to the membrane surfaces, three sophisticated models of biomembranes have been designed and subjected to specular- and off-specular X-ray and neutron scattering experiments. In order to understand the influence of molecular complexities, a variety of glycolipid molecules have been investigated, starting from simple synthetic glycolipids to complex lipopolysaccharides purified from bacteria mutants. Within the framework of this thesis, E. Schneck has successfully developed several new methods to theoretically model the measured scattering signals in a quantitative manner. The significant influence of the saccharide conformation on inter-membrane interactions and membrane mechanics suggests the crucial role of chemical structures of saccharide headgroups and mono- and divalent ions in fine-adjustment of interactions at biological interfaces. In order to obtain the amount and location of individual ion species, grazing-incidence Xray fluorescence (GIXF) was applied to complex biological membranes for the first time, demonstrating a clear displacement of monovalent ions in the core saccharide region by divalent ions. To conclude, the thesis work includes many interesting results obtained by the combinations of unique X-ray and neutron scattering techniques and thorough theoretical interpretations, which opens a new direction in physics of soft interfaces in nature.

Heidelberg, September 2010   Prof. Dr. Motomu Tanaka

Acknowledgments

I would like to thank

Prof. Dr. Motomu Tanaka for scientific guidance and many opportunities,

Prof. Dr. Annemarie Pucci for acting as my Ph.D. supervisor,

Dr. Florian Rehfeldt for constant advice and support,

Dr. Thomas Schubert for continuous scientific exchange and fruitful discussions,

Dr. Peter Seitz for lots of scientific and practical help,

Dr. Rafael Oliveira, **Dr. Fernanda Rossetti**, and **Thomas Kaindl** for help with experiments,

Dr. Bruno Demé and **Dr. Oleg Konovalov** for guidance during scattering experiments,

Prof. Dr. David Pink and **Bonnie Quinn** for computer simulations,

Dr. Christian Gege for the fancy molecules,

Dr. Stefan Kaufmann for the squash lessons, and everybody else for the great time I had as a member of the Tanaka Group.

Contents

Chapter 1
Introduction

The cells of all living organisms are confined by biomembranes which are based on self-assembled lipid bilayers. Among their various functions, membranes regulate the transport of nutrients and act as platforms for a diversity of metabolic functions. Together with the cytoskeleton, membranes are also responsible for the shape and mechanical stability of cells. The contact between the plasma membranes of neighboring cells as well as between cells and their surrounding is usually mediated by hydrated biopolymers such as extracellular matrix [1] (ECM) and glycocalix [2], both comprising various membrane-bound and membrane-associated oligo- and polysaccharides. ECM is expressed in the extracellular space and consists mainly of polysaccharide chains (e.g., glycosaminoglycans, cellulose) and fibrous proteins (e.g., collagen, laminin). Glycocalix, which covers cellular plasma membranes, is composed of membrane-bound oligo- and polysaccharide chains, chemically bound either to membrane lipids (glycolipids) or to membrane proteins (glycoproteins). Figure 1.1 (left) shows an electron micrograph of neighboring plant cells interacting via a thin layer of hydrated saccharides. As shown schematically in Fig. 1.1 (right), such cell–cell contacts can be generalized as membrane-membrane interactions, mediated via saccharide-based biopolymers.

Membrane-bound saccharides can mediate membrane-membrane interactions both in a non-specific as well as a specific manner. Non-specifically, via relatively weak (generic) forces like electrostatic interactions, hydrogen bonding, long-range van der Waals interactions, hydration- and polymer-induced forces, these saccharides act as "repellers" that maintain a finite distance between neighboring cells. Moreover, highly hydrated polysaccharides create hydrodynamic pathways for the transport of ions and molecules [3]. On the other hand, membrane-bound saccharides are also involved in specific recognition processes that are important in immune response and receptor-mediated signal transduction [4]. Although most known receptors for cell-surface carbohydrates are proteins (e.g., lectins [5, 6]), several studies have postulated that specific carbohydrate-carbohydrate interactions can be formed between complementary carbohydrate motifs [7, 8].

E. Schneck, *Generic and Specific Roles of Saccharides at Cell and Bacteria Surfaces,* Springer Theses, DOI: 10.1007/978-3-642-15450-8_1,
© Springer-Verlag Berlin Heidelberg 2011

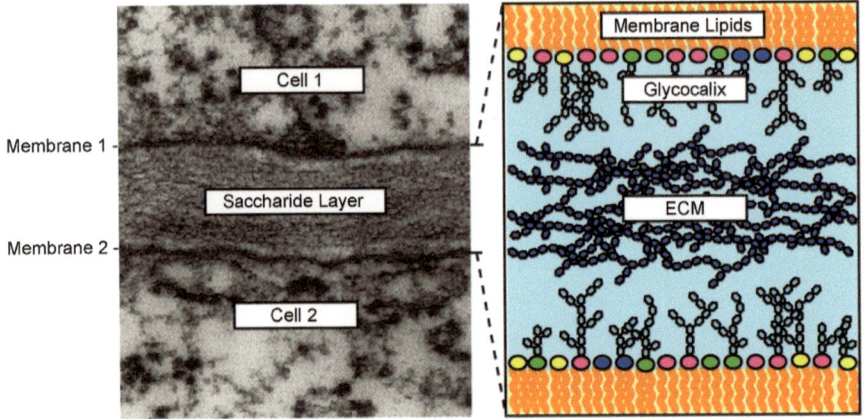

Fig. 1.1 (*Left*) electron micrograph (from: Alberts et al. [3]) of neighboring plant cells interacting via an extracellular saccharide layer. (*Right*) schematic illustration of a saccharide-mediated cell–cell contact: The ECM fills the intervening space between the membrane-bound glycocalix

 For example, the homophilic interactions (i.e., between two identical partners) of membrane-bound saccharides bearing the neutral LewisX trisaccharide motif were reported to induce the aggregation of cells [9, 10]. Such carbohydrate interactions also play an important role in cell adhesion processes during embryonic development [10].

Membrane-bound saccharides expressed on Gram-negative bacteria also play a crucial role in protecting the bacteria from their environment. Figure 1.2 (left) shows an electron micrograph of the surface of a Gram-negative bacterium. A schematic illustration is presented on the right side of the figure. The surface of the outer membrane is decorated with negatively charged saccharides, formed by lipopolysaccharides (LPSs) [12]. Beyond their structural role in the outer membrane leaflet, LPSs can serve as a barrier against harmful molecules. For instance,

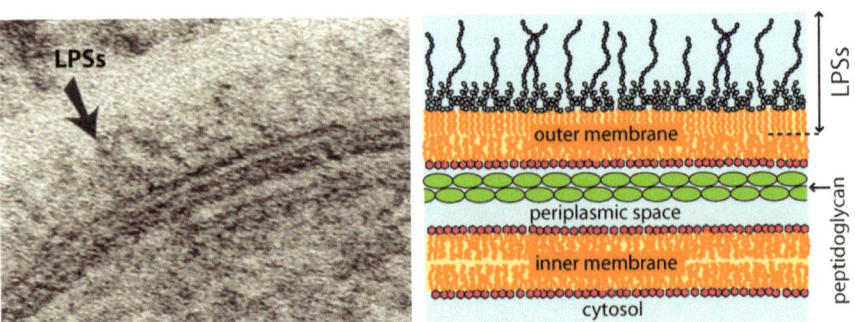

Fig. 1.2 (*Left*) electron micrograph (from: Beveridge et al. [11]) of a Gram-negative bacteria surface. (*Right*) architecture of the saccharide-rendered double membrane of Gram-negative bacteria. Saccharide units are indicated with hexagons

LPSs are implicated in the resistance of Gram-negative bacteria against cationic antimicrobial peptides (CAPs) in the presence of divalent cations (e.g., Ca^{2+}, Mg^{2+}). Since the first report of Brock [13], many in vivo studies [14, 15] have demonstrated that divalent cations significantly increase the minimum inhibitory concentration (MIC), i.e., the concentration of CAPs required to inhibit bacterial growth. The presence of 10 mM $MgCl_2$ was found to increase the MIC of *Pseudomonas aeruginosa* PAO1 against protamine, a CAP from sperm cells of vertebrates, by a factor of five [16]. The detailed investigation of this phenomenon is of fundamental importance for the understanding of the function of antimicrobial peptides as well as for the development of peptide-based antibiotics [17].

There have also been several theoretical approaches to explain this effect of divalent cations on bacterial resistance against CAPs. Atomic-scale molecular dynamic (MD) simulations [18–20] suggested that divalent cations condense in the negatively charged "core region" of the LPS molecules. However, the dimensions of the simulation box (a few nm) and the time scales (several ns) were insufficient to monitor conformational changes of the saccharide chains of LPS molecules. Recently, Pink et al. [16]. introduced a coarse-grained "minimum computer model" of LPS surfaces that allows for the Monte Carlo (MC) simulation of larger volumes (more than 100 LPS molecules) over longer time scales (in the order of ms). This approach seems generally suited to model the conformation of LPS molecules in the presence and absence of divalent cations as well as the electrostatics of bacteria surfaces rendered with charged saccharide chains. However, experimental studies that reveal the detailed structure of such soft, complex interfaces under biologically relevant conditions are still missing.

In this thesis, three classes of planar models (Fig. 1.3) of saccharide-rendered cell and bacteria surfaces were prepared using well-defined molecular building blocks: multilayers of glycolipid membranes supported by solid substrates (Fig. 1.3a), glycolipid monolayers deposited on hydrophobic substrates (Fig. 1.3b), and glycolipid monolayers at the air/water interface (Fig. 1.3c). By taking advantage of the planar sample geometry, the in-plane and out-of-plane structures of these model systems were investigated using a variety of X-ray and neutron scattering techniques.

In contrast to other high-resolution probing methods (e.g., scanning tunneling microscopy, atomic force microscopy, and electron microscopy), scattering techniques do not require direct access to the surfaces, but can also reveal buried structures. Since the 1970s, small and wide-angle X-ray scattering (SAXS and WAXS) have commonly been used for the physical characterization of phospholipid membranes suspensions [21–33]. To date, several groups have used these methods to determine the structures of isotropic glycolipid membrane suspensions [34, 35], the influence of small water-soluble carbohydrates on phospholipid membranes [36], the influence of lacitol glycolipids on charged membranes [37], and the structure of membranes incorporating ganglioside (GM1) molecules [38]. The same approach could also be used to study suspensions of more complex molecules, such as bacterial lipopolysaccharides [39–43]. However, the random orientation of membranes in suspensions generally does not allow for

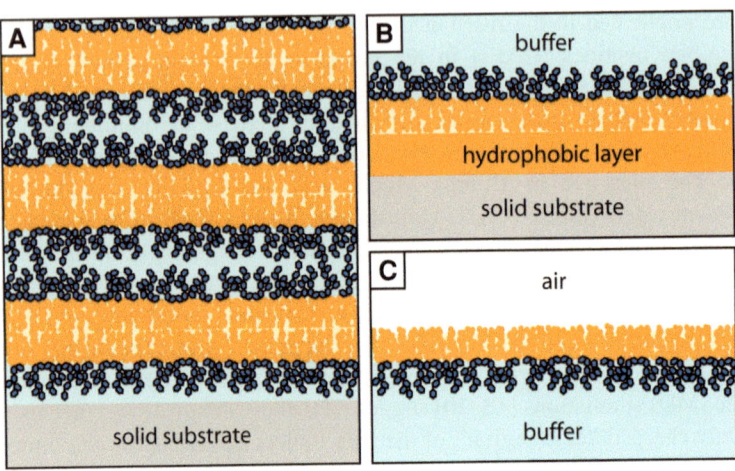

Fig. 1.3 *Planar* oriented models of saccharide-rendered cell surfaces used in this study. **a** Solid-supported glycolipid membrane multilayers. **b** Solid-supported glycolipid monolayers. **c** Glyco-lipid monolayers at the air/water interface. The model systems are studied using various X-ray and neutron scattering techniques

the distinct determination of structures perpendicular and parallel to the membrane planes. This problem can be overcome by the use of planar model systems (see Fig. 1.3), where specular and off-specular scattering signals can be identified [23, 24, 32, 33]. Information on the structure normal to the sample plane can be obtained from specular scattering, whereas information on the structural ordering parallel to the sample plane, reflecting the mechanical properties of the membranes, can be extracted from off-specular signals. To date, this approach has mainly been used for synthetic lipid membranes [21, 27, 28, 31], but also for mixtures of lipids with peptides/proteins [44, 45] or cholesterol [46]. Although several recent studies dealt with LPS membrane multilayers on solid substrates, they did not take advantage of the planar geometry, but solely focused on lamellar periodicities and out-of-plane membrane structures under various conditions [47–49]. Studies that reveal both in-plane and out-of plane structures of saccharide-rendered cell and bacteria surfaces under biological conditions are still missing.

In the present work, planar membrane models (Fig. 1.3) are utilized for a systematic study of membrane-bound saccharides by various X-ray and neutron scattering techniques. In Chap. 5 the influence of membrane-bound saccharides on the interactions and mechanical properties of membranes are investigated by specular and off-specular neutron scattering from synthetic glycolipid membrane stacks. To highlight the role of the saccharide conformation, two types of glyco-lipids with a distinct difference in their disaccharide head group are compared (Sect. 5.1). In Sect. 5.2, synthetic glycolipids bearing the LewisX trisaccharide motif were incorporated into membrane multilayers to study the influence of specific saccharide-saccharide interactions on multilayer structure and mechanics.

In Chap. 6, the focus is extended towards more complex models of bacterial surfaces, prepared from lipopolysaccharides of various structural and compositional complexities, ranging from monodisperse rough mutant LPS molecules to native polydisperse LPS extracts from *Pseudomonas aeruginosa* [50]. Here, in particular the influence of divalent cations on the LPS conformation and on the mechanics of LPS membranes is investigated. The samples are studied using neutron scattering, high-energy X-ray reflectometry, and grazing-incidence X-ray fluorescence (GIXF). In Chap. 4, the theoretical methods developed in this thesis for a quantitative interpretation of specular and off-specular neutron scattering (Sect. 4.1) and X-ray fluorescence signals (Sect. 4.3) are derived and discussed.

References

1. W.D. Comper, *Extracellular Matrix* (Harwood Academic Publishers, Amsterdam, 1996)
2. H.J. Gabius, S. Gabius, *Glycoscience* (Chapmann & Hall, Weinheim, 1997)
3. B. Alberts, D. Bray, J. Lewis, M. Raff, K. Roberts, J.D. Watson, *Molecular Biology of the Cell* (Garland Science, New York, 2002)
4. D. Voet, J.G. Voet, *Biochemistry* (Wiley, NY, 1995)
5. C.J. Nilsson, *Lectins: Analytical Technologies* (Elsevier, Amsterdam, 2007)
6. N. Sharon, H. Lis, *Lectins* (Kluwer, Dordrecht, 2003)
7. N. Seah, A. Basu, *Encyclopedia of Chemical Biology* (Wiley, NY, 2008)
8. C. Tromas, J. Rojo, J.M. de la Fuente, A.G. Barrientos, R. García, S. Penadés, Adhesion forces between Lewis X determinant antigens as measured by atomic force microscopy. Angew. Chem. Int. Ed. **40**, 3052 (2001)
9. M. Boubelik, D. Floryk, J. Bohata, L. Draberova, J. Macak, F. Smid, P. Draber, LeX glycosphingolipid mediated cell aggregation. Glycobiology **8**, 139 (1998)
10. I. Eggens, B. Fenderson, T. Toyokuni, B. Dean, M. Stroud, S. Hakomori, Specific interaction between Lex and Lex determinants. A possible basis for cell recognition in preimplantation embryos and in embryonal carcinoma cells. J. Biol. Chem. **264**, 9476 (1989)
11. T.J. Beveridge, L.L. Graham, Surface Layers of Bacteria. Microbiol. Rev. **55**, 684 (1991)
12. O. Lüderitz, M. Freudenberg, C. Galanos, V. Lehmann, E.T. Rietschel, D.H. Shaw, Lipopolysaccharides of Gram-negative bacteria. Curr. Top. Membr. Transp. **17**, 79 (1982)
13. T.D. Brock, The effect of salmine on bacteria. Can. J. Microbiol. **4**, 65 (1958)
14. L. Truelstrup Hansen, J.W. Austin, T.A. Gill, Antibacterial effect of protamine in combination with EDTA and refrigeration. Int. J. Food Microbiol. **66**, 149 (2001)
15. N.M. Islam, T. Itakura, T. Motohiro, Antibacterial characteristics of fish protamines. 1: Antibacterial spectra and minimum inhibition concentration of clupeine and salmine. Bull. Jpn. Soc. Sci. Fish. **50**, 1705 (1984)
16. D.A. Pink, L.T. Hansen, T.A. Gill, B.E. Quinn, M.H. Jericho, T.J. Beveridge, Divalent calcium ions inhibit the penetration of protamine through the polysaccharide brush of the outer membrane of Gram-negative bacteria. Langmuir **19**, 8852 (2003)
17. R.E.W. Hancock, D.S. Chapple, Peptide antibiotics. Antimicrob. Agents Chemother. **43**, 1317 (1999)
18. L.P. Kotra, D. Golemi, N.A. Amro, G.Y. Liu, S. Mobashery, Dynamics of the lipopolysaccharide assembly on the surface of *Escherichia coli*. J. Am. Chem. Soc. **121**, 8707 (1999)
19. R.D. Lins, T.P. Straatsma, Computer simulation of the rough lipopolysaccharide membrane of *Pseudomonas aeruginosa*. Biophys. J. **81**, 1037 (2001)

20. R.M. Shroll, T.P. Straatsma, Molecular structure of the outer bacterial membrane of *Pseudomonas aeruginosa* via classical simulation. Biopolymers **65**, 395 (2002)
21. G. Brotons, L. Belloni, T. Zemb, T. Salditt, Elasticity of fluctuating charged membranes probed by X-ray grazing-incidence diffuse scattering. Europhys. Lett. **75**, 992 (2006)
22. T.A. Harroun, M. Koslowsky, M.-P. Nieh, C.-F. de Lannoy, V.A. Raghunathan, J. Katsaras, Comprehensive examination of mesophases formed by DMPC and DHPC mixtures. Langmuir **21**, 5356 (2005)
23. N. Lei, C.R. Safinya, R.F. Bruinsma, Discrete harmonic model for stacked membranes—theory and experiment. J. Phys. II **5**, 1155 (1995)
24. E.A.L. Mol, J.D. Shindler, A.N. Shalaginov, W.H. de Jeu, Correlations in the thermal fluctuations of free-standing smectic-A films as measured by X-ray scattering. Phys. Rev. E **54**, 536 (1996)
25. G. Pabst, H. Amenitsch, D.P. Kharakoz, P. Laggner, M. Rappolt, Structure and fluctuations of phosphatidylcholines in the vicinity of the main phase transition. Phys. Rev. E **70**, 021908 (2004)
26. G. Pabst, M. Rappolt, H. Amenitsch, P. Laggner, Structural information from multilamellar liposomes at full hydration: full q-range fitting with high quality X-ray data. Phys. Rev. E **62**, 4000 (2000)
27. B. Pozo-Navas, V.A. Raghunathan, J. Katsaras, M. Rappolt, K. Lohner, G. Pabst, Discontinuous unbinding of lipid multibilayers. Phys. Rev. Lett. **91**, 02101 (2003)
28. M.C. Rheinstädter, C. Ollinger, G. Fragneto, T. Salditt, Collective dynamics of lipid membranes studied by inelastic neutron scattering. Phys. Rev. Lett. **93**, 108107 (2004)
29. C.R. Safinya, D. Roux, G.S. Smith, S.K. Sinha, P. Dimon, N.A. Clark, A.M. Bellocq, Steric interactions in a model multimembrane system: a synchrotron X-ray study. Phys. Rev. Lett. **57**, 2718 (1986)
30. C.R. Safinya, E.B. Sirota, D. Roux, G.S. Smith, Universality in interacting membranes—the effect of cosurfactants on the interfacial rigidity. Phys. Rev. Lett. **62**, 1134 (1989)
31. T. Salditt, Thermal fluctuations and stability of solid-supported lipid membranes. J. Phys. Cond. Matt. **17**, R287 (2005)
32. L. Yang, T.A. Harroun, W.T. Heller, T.M. Weiss, H.W. Huang, Neutron off-plane scattering of aligned membranes. I. Method of measurement. Biophys. J. **75**, 641 (1998)
33. Y. Lyatskaya, Y. Liu, S. Tristram-Nagle, J. Katsaras, J.F. Nagle, Method for obtaining structure and interactions from oriented lipid bilayers. Phys. Rev. E **63**, 011907 (2000)
34. M.F. Schneider, R. Zantl, C. Gege, R.R. Schmidt, M. Rappolt, M. Tanaka, Hydrophilic/hydrophobic balance determines morphology of glycolipids with oligolactose headgroups. Biophys. J. **84**, 306 (2003)
35. M. Tanaka, M.F. Schneider, G. Brezesinski, In-plane structures of synthetic oligolactose lipid monolayers—impact of saccharide chain length. Chem. Phys. Chem. **4**, 1316 (2003)
36. B. Demé, M. Dubois, T. Zemb, B. Cabane, Effect of carbohydrates on the swelling of a lyotropic lamellar phase. J. Phys. Chem. **100**, 3828 (1996)
37. F. Ricoul, M. Dubois, L. Belloni, T. Zemb, C. Andr-Barrs, I. Rico-Lattes, Phase equilibria and equation of state of a mixed cationic surfactant-glycolipid lamellar system. Langmuir **14**, 2645 (1998)
38. T.J. McIntosh, S.A. Simon, Long- and short-range interactions between phospholipid/ganglioside GM1 bilayers. Biochemistry **33**, 10477 (1994)
39. K. Brandenburg, U. Seydel, Physical aspects of structure and function of membranes made from lipopolysaccharides and free lipid A. Biochim. Biophys. Acta **775**, 225 (1984)
40. M. Kastowsky, T. Gutberlet, H. Bradaczek, Comparison of X-ray powder-diffraction data of various bacterial lipopolysaccharide structures with theoretical model conformations. Eur. J. Biochem. **217**, 771 (1993)
41. N. Kato, M. Ohta, N. Kido, H. Ito, S. Naito, T. Hasegawa, T. Watabe, K. Sasaki, Crystallization of R-form lipopolysaccharides from *Salmonella minnesota* and *Escherichia coli*. J. Bacteriol. **172**, 1516 (1990)

42. H. Labischinski, G. Barnickel, H. Bradaczek, D. Naumann, E.T. Rietschel, P. Giesbrecht, High state of order of isolated bacterial lipopolysaccharide and its possible contribution to the permeation barrier property of the outer membrane. J. Bacteriol. **162**, 9 (1985)
43. U. Seydel, K. Brandenburg, M.H.J. Koch, E.T. Rietschel, Supramolecular structure of lipopolysaccharide and free lipid A under physiological conditions as determined by synchrotron small-angle X-ray diffraction. Eur. J. Biochem. **186**, 325 (1989)
44. T. Salditt, Lipid–peptide interaction in oriented bilayers probed by interface-sensitive scattering methods. Curr. Opinion Struct. Biol. **13**, 467 (2003)
45. T. Salditt, G. Brotons, Biomolecular and amphiphilic films probed by surface sensitive X-ray and neutron scattering. Anal. Bioanal. Chem. **379**, 960 (2004)
46. J. Pan, T.T. Mills, S. Tristram-Nagle, J.F. Nagle, Cholesterol perturbs lipid bilayers nonuniversally. Phys. Rev. Lett. **100**, 198103 (2008)
47. T. Abraham, S.R. Schooling, M. Nieh, N. Kucerka, T.J. Beveridge, J. Katsaras, Neutron diffraction study of *Pseudomonas aeruginosa* lipopolysaccharide bilayers. J. Phys. Chem. B **111**, 2477 (2007)
48. N. Kucerka, E. Papp-Szabo, M. Nieh, T.A. Harroun, S.R. Schooling, J. Pencer, E.A. Nicholson, T.J. Beveridge, J. Katsaras, Effect of cations on the structure of bilayers formed by lipopolysaccharides isolated from *Pseudomonas aeruginosa* PAO1. J. Phys. Chem. B **112**, 8057 (2008)
49. S. Snyder, D. Kim, T.J. McIntosh, Lipopolysaccharide bilayer structure: effect of chemotype, core mutations, divalent cations, and temperature. Biochemistry **38**, 10758 (1999)
50. N. Høiby, Prevention and treatment of infections in cystic fibrosis. Int. J. Antimicrob. Agents **1**, 229 (1992)

Chapter 2
Theoretical Background

In this chapter the theoretical concepts and physical principles used in this thesis are presented. The first part deals with the physical description of the studied systems, while the second part provides an introduction into the principles of X-ray and neutron scattering, the primary experimental method applied in this thesis.

2.1 Lipid Membranes

2.1.1 Physics of Lipids and Lipid Membranes

In this thesis, membrane-bound saccharides are studied using model systems prepared from various types of lipids. In the following, an introduction into the physics of lipids and lipid membranes is given.

2.1.1.1 Membrane Formation by Lipid Self-Assembling

Lipids are the main structuring component of biological membranes. Despite their enormous structural variety, all membrane lipids possess amphiphilic architecture, with one hydrophobic and one hydrophilic building block, essential for the molecular self-assembling.[1] The hydrophobic part is constituted by apolar hydrocarbon chains, while the hydrophilic part, commonly called head group, consists of polar moieties. This is illustrated in Fig. 2.1 for DPPC, a commonly studied lipid with two hydrocarbon chains and a phosphatidylcholine head group.

---

[1] Archaea possess lipids with two hydrophilic moieties, which leads to different molecular self-assembly.

E. Schneck, *Generic and Specific Roles of Saccharides at Cell and Bacteria Surfaces,* Springer Theses, DOI: 10.1007/978-3-642-15450-8_2,
© Springer-Verlag Berlin Heidelberg 2011

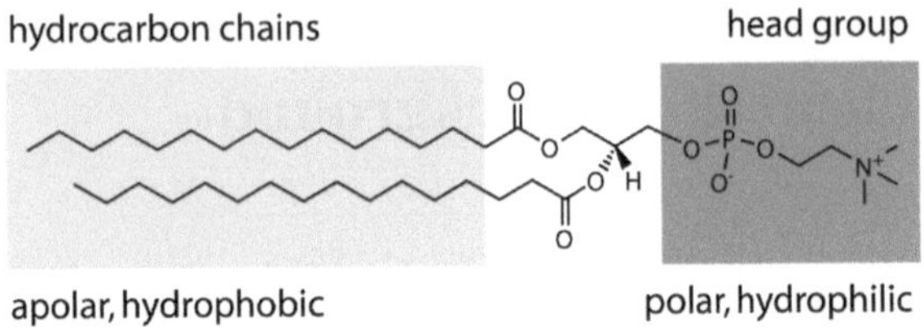

Fig. 2.1 Schematic structure of a lipid. The amphiphilic molecules possess a hydrophobic part, constituted by apolar hydrocarbon chains, and a hydrophilic head group exhibiting polar moieties

hydrocarbon chains   head group

apolar, hydrophobic   polar, hydrophilic

Due to the hydrophobic effect [1, 2], lipids exhibit a very low solubility in water. Isolated lipids dissolved in the aqueous solution are only found in very low concentrations, while the vast majority of lipids form aggregates whose shapes depend on the lipid geometry characterized by the lipid packing parameter [2]. Here, we focus on the biologically most relevant cylindrical lipid geometry, which leads to the formation of bilayer structures (see Fig. 2.2). In the following, the self-assembling of lipids is discussed under thermodynamic aspects. Biological processes usually take place at constant temperature T and constant (atmospheric or hydrostatic) pressure. Under these boundary conditions, the processes are driven by a minimization of the Gibbs free energy G.

$$G = H - TS,$$

where H denotes the enthalpy and S the entropy. The solubility of the lipids can be quantified in terms of the critical aggregate concentration (CAC), which defines an upper limit for the concentration of isolated lipids dissolved in the aqueous solution [1, 3].

$$\text{CAC} \cong c_0 \exp(-\Delta g / k_B T)$$

$$\Delta g = g_{\text{sol}} - g_{\text{agg}}$$

Here, $k_B T$ denotes the thermal energy and Δg the change in the Gibbs free energy that would result from the transfer of one lipid from an aggregate into the solution. c_0 denotes the concentration of lipids within the aggregates, usually at

Fig. 2.2 Bilayer formation due to the hydrophobic effect. Isolated lipids dissolved in the aqueous solution are only found in very low concentration below the critical aggregate concentration (CAC), while the vast majority of the lipids are assembled in aggregates such as the shown lipid bilayer

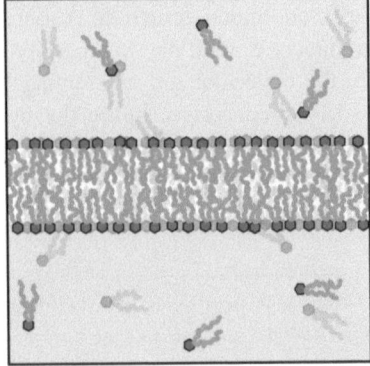

the order of 1 M. Typical lipids with two hydrocarbon chains have CAC values at the order of 10^{-12} M. Δg can be decomposed into an enthalpic contribution Δh and an entropic contribution $T\Delta s$:

$$\Delta g = \Delta h - T\Delta s$$

Δh is determined by the enthalpy of lipid/lipid interactions, water/water interactions, and lipid/water interactions. Δs is determined by the change in the number of available system configurations that would result from the transfer of one lipid into solution. A dissolved lipid exposes its apolar hydrocarbon chains to the adjacent water molecules, which thus get limited in the choice of H-bonds they can form. Thus, the number of available system configurations gets reduced if a lipid is brought into solution. As a consequence, Δs is always negative. Usually the absolute of $T\Delta s$ is much higher than the absolute of Δh, and Δg is dominated by the entropic contribution which always favors the formation of aggregates. This phenomenon, known as hydrophobic effect, is the driving force for the self-assembling of lipids into aggregates, such as bilayer membranes (see Fig. 2.2) or other structures.

2.1.1.2 Phase Behavior of Lipid Membranes

Depending on the lipid structure, lipid membranes can assume several phase states in many cases. These phase states are usually classified as follows [4, 5]: The L_c-phase represents a two-dimensional crystalline arrangement of the lipid molecules including hydrocarbon chains and head groups. The gel-like L_β-phase represents a two-dimensional crystalline ordering of the hydrocarbon chains, while the headgroups have no fixed orientation or position. The hydrocarbon chains are arranged in a lateral hexagonal-like (generally "orthorhombic") structure with a fixed, linear configuration, denoted as all-trans configuration. This is illustrated in Fig. 2.3 (top membrane). Often the chains posses a defined tilt with respect to the membrane normal, due to a mismatch between the projected area required by the hydrocarbon chains and that required by the head groups.[2] The biologically most relevant L_α-phase represents a two-dimensional fluid, where the hydrocarbon chains can assume a large number of configurations (trans and gauche rotations at each carbon–carbon bond in the hydrocarbon chains) [6], and the lipids diffuse along the membrane plane. This is illustrated in Fig. 2.3 (bottom membrane). Due to the high number of gauche configurations, membranes are significantly thinner in fluid L_α-phase than in gel-like L_β-phase [3].

Lipid bilayers can transit from one phase state to another one via thermotropic phase transitions.[3] Here, which phase state is assumed by the bilayer depends on

---

[2] This is the case for lipids with phosphatidylcholine headgroups (such as DPPC), where this tilted phase is denoted as $L_\beta{}'$-phase.

[3] Phase transitions can also be induced by other thermodynamic parameters, such as pressure or concentration.

Fig. 2.3 A lipid membrane in gel-like L_β-phase and in fluid L_α-phase

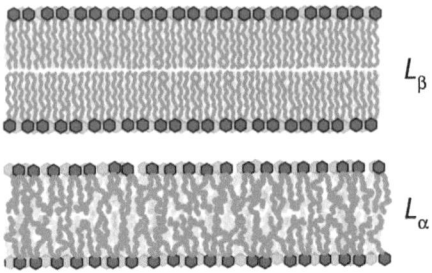

the temperature T, while other parameters (lipid concentration, pressure) are kept constant. In the following we consider the biologically very important thermotropic phase transition from L_β-phase to L_α-phase. This first order phase transition is known as chain-melting transition. Let Δg denote the difference in Gibbs free energy per lipid between the two phase states for a given temperature T.

$$g_\alpha - g_\beta = \Delta g = \Delta h - T\Delta s$$

Δh has a positive value, as the In L_α-phase gauche configurations with higher energy levels are occupied, while in the L_β-phase, the hydrocarbon chains mainly take trans configurations with lowest energy. Hence, the enthalpic contribution of the Gibbs free energy favors the L_β-phase. On the other hand, the large number of possible chain configurations in L_α-phase [6] coincides with a significantly higher entropy per lipid and results in a positive value of Δs. Thus, the entropic contribution of the Gibbs free energy favors the L_α-phase. As a consequence, the temperature determines which contribution is dominant in this competition. At the phase transition temperature T_m, where the lipids are found in either phase with the same probability, Δg is zero, and thus:

$$T_m = \Lambda h / \Delta s$$

The transition temperature is determined by the transition enthalpy Δh and the transition entropy Δs, which are both encoded in the molecular structure of the lipids. Conversely, measurements of T_m provide valuable information on the thermodynamics of the bilayers formed by the studied lipids. Below T_m bilayers assume L_β-phase and above T_m they assume L_α-phase. Near T_m, a coexistence of the two phases is found. The width of this coexistence region depends on the transition cooperativity, i.e., on the size of the group of lipids that undergo the transition in a cooperative process [5].

2.1.2 Inter-Membrane Interactions

In this thesis, the influence of membrane-bound saccharides on the interaction of membranes is investigated. In the following, an introduction into the physics of inter-membrane interactions is given.

In general, the physical interaction between biological membranes is the result of a complex interplay of non-specific and specific interactions. The former comprise a variety of generic physical forces, while the latter correspond to specific bonds between motifs (e.g. oligosaccharides) presented by the membranes and can be treated separately (see Sect. 5.2). For large enough membrane separations (typically $\gg 10$ Å), the non-specific interactions can be treated in a continuum mean-field approximation which neglects the atomic structure of the matter. Here, the bilayers are considered as plates with in-plane-homogenous properties (such as charge density, dielectric constant, and bending rigidity). This is a reasonable assumption for many model membranes. Commonly used continuum mean-field approaches to describe the interaction of particles in aqueous media are the Gouy–Chapman "diffuse double layer" theory, which accounts for electrostatic interactions in electrolytes, and the DLVO theory [1, 7–10], which additionally accounts for the van der Waals interaction. However, for the description of the interaction between planar membranes, additional contributions, not covered by the above mentioned theories, become important, like hydration or undulation forces.

2.1.2.1 Hardcore Repulsion

Hardcore repulsion between membranes is important for vanishing inter-membrane separations, where the continuum description of most of the other force contributions fails. However, once hardcore repulsion becomes relevant, it dominates over all the other contributions. Hardcore repulsion originates from an overlap of orbitals of the adjacent membrane molecules. Empirically, the resulting potential can be described with a power law:

$$\Pi_{HC}(d_W) = d_W^{-(n+1)}, \quad 9 < n < 16$$

Even though the theoretical basis for this description is very poor, it is commonly used because of its mathematical convenience [1, 11].

2.1.2.2 Van der Waals Interaction

The van der Waals interaction accounts for the combination of three contributions [1]:

1. The interaction arising from the orientation of permanent dipoles, known as orientation force.
2. The interaction between permanent dipoles with induced dipoles, known as induction force.
3. The interaction between induced dipoles with induced dipoles, which is always present and known as dispersion force.

The Lifshitz theory describes the van der Waals interaction between homogenous media in a continuum approximation. This description is based on the calculation of Hamaker constants A, which depend on the dielectric properties of the interacting media [1, 12, 13]. The interaction can be attractive or repulsive, in general. However, in case of two identical media (here: lipid membranes) with the same thickness d_H, separated by a third medium (here: water) of thickness d_W, the interaction is always attractive. For lipid membranes several models have been developed [14, 15]. As demonstrated by Demé et al. [16], a double film model with one Hamaker constant $A = 5.1 \times 10^{-21}$ J is sufficient to account for the van der Waals attraction between phospholipid membranes in fluid L_α-phase:

$$\Pi_{VDW}(d_W) = -\frac{A}{6\pi}\left[\frac{1}{d_W^3} - \frac{2}{(d_W + d_H)^3} + \frac{1}{(d_W + 2d_H)^3}\right]$$

Here, the membrane separation d_W is defined as the thickness of the aqueous region separating the hydrophobic membrane regions characterized by the "hydrophobic thickness" d_H.

2.1.2.3 Hydration Repulsion

Hydration repulsion between hydrophilic surfaces is believed to result from the free energy required to modify the H-bonding network of liquid water in the vicinity of polar, H-bonding surface groups. Empirically, the hydration interaction obeys an exponential decay [1] characterized by the extrapolated pressure Π_0 at $d_W = 0$ and a decay length λ_{HYD}:

$$\Pi_{HYD}(d_W) = \Pi_0 \exp\left(-d_W/\lambda_{HYD}\right)$$

For interacting phosphatidylcholine lipid membranes in fluid L_α-phase, $\Pi_0 = 4.5 \times 10^9$ Pa is an established value, if the membrane separation d_W is defined in the same way as for the van der Waals interaction [16]. For λ_{HYD}, values around 2.0 Å were reported [17, 18, 53].

2.1.2.4 Undulation Repulsion

Up to here it was sufficient to treat the interacting lipid membranes as perfectly flat layers. However, the undulation repulsion can only be understood, if thermal membrane fluctuations are taken into account. These out-of-plane "undulations" depend on the membrane bending rigidity κ. The undulation repulsion between two membranes originates from the suppression of undulation modes (which represents a decrease in entropy) as the membrane separation d_W decreases. The strength of the repulsion depends on the geometry in which the membranes are confined. For a stack of N membranes, an algebraic expression was derived by Helfrich [19–21],

$$\Pi_{\text{UND}}(d_{\text{W}}) = \frac{2N}{N+1} \alpha_N \frac{(k_B T)^2}{\kappa d_{\text{W}}^3}.$$

Here, the membrane separation d_{W} is defined as the thickness of the aqueous region between the "steric surfaces" of the adjacent membranes. In case of solid-supported membrane multilayers, where N is large enough, this can be approximated with the limit of $N \to \infty$:

$$\Pi_{\text{UND}}(d_{\text{W}}) = \alpha_\infty \frac{2(k_B T)^2}{\kappa d_{\text{W}}^3},$$

with the analytic estimate $\alpha_\infty = 3\pi^2/128 \approx 0.23$ provided by Helfrich. More recent calculations by Bachmann et al. [19] indicate a lower value of $\alpha_\infty \approx 0.104$. Generally, the undulation repulsion becomes important for larger separations, where the membranes are coupled only weakly by the other force contributions.

2.1.2.5 Electrostatic Interaction

Electrostatic interactions occur if the interacting membranes carry electric charges. This can be due to charged headgroup residues or due to the adsorption of ions to the membrane surface (see Sect. 5.2.2.2). In membrane multilayers with homogenous molecular composition, the membranes are like-charged and the electrostatic interaction is therefore repulsive. In a continuum approximation the in-plane distribution of the elementary charges is neglected, and a homogenous surface charge density σ is assumed, which is confined in a defined plane of the membrane. The electrostatic interaction between adjacent membranes across the aqueous medium depends in a complicated way on the surface charge and on the ion concentrations of the buffer solution. Since there are no general algebraic solutions available, this problem is further developed in Sect. 4.2.

2.1.2.6 The Disjoining Pressure

The disjoining pressure constitutes the net force (per unit area) resulting from all interactions n between two planes (here: membrane surfaces) and is defined as the negative derivative of the Gibbs free energy with respect to the plane separation d_{W} while the chemical potential μ for water is kept constant:

$$\Pi = \sum_n \Pi_n = \left(\frac{\partial G}{\partial d_{\text{W}}} \right)_\mu$$

A stable equilibrium separation is found wherever:

$$\Pi = 0 \quad \text{and} \quad \frac{\partial \Pi}{\partial d_{\text{W}}} < 0$$

In contrast to the individual force contributions Π_n, the disjoining pressure is accessible to experimental measurements. If sufficient information about the interacting membranes is available, the experimentally determined disjoining pressure profile $\Pi(d_W)$ can be represented theoretically with a superposition of the relevant force contributions.

2.1.2.7 Force–Distance Relationships

The disjoining pressure profile can be determined in measurements of the membrane separation d_W while various well-known external compressional forces Π_{ext} are exerted to the membranes. In the presence of an external force, the equilibrium membrane separation is shifted to the value, where $\Pi + \Pi_{ext} = 0$. This yields the identity $\Pi(d_W) = -\Pi_{ext}(d_W)$.

Typically d_W is determined indirectly by measuring the lamellar periodicity d of membrane multilayers and subsequent subtraction of the membrane thickness. The so recorded data points provide quantitative relationships between the membrane separation and the disjoining pressure, referred to as force–distance relationships. Low compressive forces ($\Pi < 10^6$ Pa) can be applied by adding impermeable solutes (e.g. water soluble polymers impermeable across the membrane) to the aqueous solution [11, 22]. This approach is known as osmotic stress method. On the other hand, measurements at controlled relative humidity h_{rel} (i.e., in the absence of condensed water) allow for the application of high compressive forces ($\Pi > 10^6$ Pa) [23]:

$$\Pi_{osm} = -\frac{k_B T}{V_{water}} \ln(h_{rel})$$

In practice, force–distance relationships are commonly presented as $\Pi(d)$ since this representation is equivalent and closer to the experimentally accessible lamellar periodicity d (see Chaps. 5, 6).

2.1.3 Mechanics of Solid-Supported Membrane Multilayers

In this thesis, the influence of membrane-bound saccharides on the mechanical properties of (interacting) membranes is studied using solid-supported membrane multilayers as model systems (see Chaps. 5, 6). In the following, an introduction into the mechanics of planar membrane multilayers is given.

2.1.3.1 The Discrete Smectic Hamiltonian

Within the framework of a continuum model approximation, the total free energy of a set of oriented membrane multilayers can be described with the discrete smectic Hamiltonian H, according to Lei et al. [54]:

$$H = \int_A d^2r \sum_{n=1}^{N-1} \left(\frac{B}{2d}(u_{n+1} - u_n)^2 + \frac{\kappa}{2}\left(\nabla_{xy}^2 u_n\right)^2 \right).$$

N is the total number of membranes, d their equilibrium periodicity, and A the area occupied by the set of multilayers. At any moment in time, each point of a membrane, defined by the in-plane coordinates x and y, is displaced from the average membrane z-position by an increment $u_n(x, y)$, due to thermal fluctuations. This is illustrated in Fig. 2.4.

Within the continuum framework, vertical compression is characterized by the compression modulus B, while bending is characterized by the membrane bending modulus κ. ∇_{xy} denotes the two-dimensional Nabla operator in x and y directions. In a harmonic approximation, B can also be expressed in terms of the derivative of the disjoining pressure with respect to the membrane separation d_W at the equilibrium lamellar periodicity d:

$$B = -d\left(\frac{\partial \Pi}{\partial d_W}\right)_d$$

The properties of the membrane fluctuations represent the underlying mechanical parameters, which are commonly characterized by the Caillé parameter η and the de Gennes parameter λ of smectic liquid crystals, as these quantities are more directly accessible to experiments (see Sect. 4.1).

$$\eta = \frac{\pi k_B T}{2d^2\sqrt{\kappa B/d}} \quad \text{Caille parameter}$$

$$\lambda = \sqrt{\frac{\kappa}{Bd}} \quad \text{de Gennes parameter}$$

For a given Temperature T, a high η value corresponds to a "soft" system and leads to strong fluctuations with a high amplitude, while a low value corresponds to a "stiff" system with weak fluctuations. On the other hand, the de Gennes parameter indicates if the stiffness of a system is rather dominated by the compression modulus (for low λ values) or rather dominated by the bending modulus (for high λ values). The de Gennes parameter also determines how strongly the fluctuations of

Fig. 2.4 Parameterization of a set of rough layers or interfaces

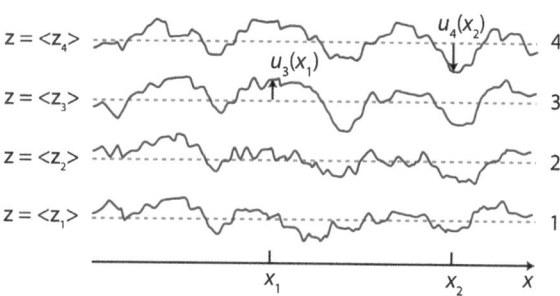

Fig. 2.5 Influence of the de
Gennes parameter λ on the
inter-membrane correlation
of fluctuations at different
wavelength

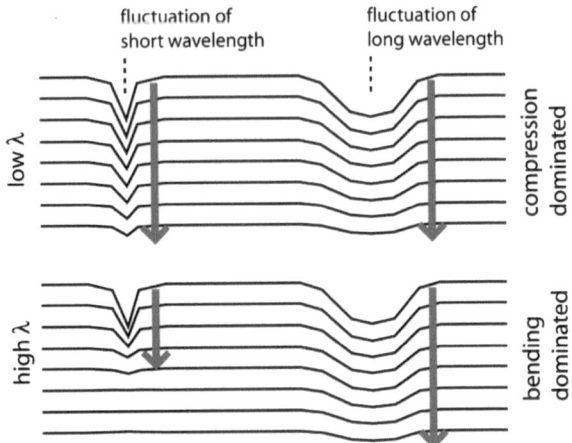

a membrane are correlated with those of its neighbors, depending on the fluctuation
wavelength. This is qualitatively illustrated in Fig. 2.5, where the local
displacements of a membrane and the resulting displacements of its neighbors
are depicted for different displacement wavelengths and de Gennes parameters.
In case of a low λ value the transmission depth of the displacement is only weakly
dependent on the wavelength, whereas in case of a high λ value a displacement of
short wavelength is damped much more strongly than that of long wavelength.
In other words, long-wavelength membrane fluctuations have a stronger
inter-membrane correlation than short wavelength fluctuations.[4]

2.1.3.2 Height–Height Correlation Functions

Correlation functions provide a model-independent statistical description of the
height profile of rough layers or interfaces (see Fig. 2.4). A single rough layer or
interface is described the height–height correlation function.

$$C(x,y) := \langle u(x_0 - x, y_0 - y) \cdot u(x_0, y_0) \rangle_{x_0, y_0}$$

The topological root mean square (rms) roughness σ of a layer or an interface is
also defined via the height–height correlation function:

$$\sigma := \sqrt{\langle u^2 \rangle} = \sqrt{C(0,0)}$$

For stratified systems, the correlation functions include self-correlation
functions ($n = m$) within a layer or an interface and cross-correlation functions
($n \neq m$) between a layer or interface and its kth neighbor.

---

[4] This is the reason why the extent, to which the width of a Bragg sheet increases with $q_{\parallel}$ (the
reciprocal space analogue to an in-plane wavelength), increases along with the de Gennes
parameter.

$$C_{nm}(x, y) := \langle u_n(x_0 - x, y_0 - y) \cdot u_m(x_0, y_0) \rangle_{x_0, y_0}$$

For laterally isotropic systems the correlation functions possess radial symmetry:

$$C(x, y) \rightarrow C(r) := \langle u(\vec{r_0} - \vec{r}) \cdot u(\vec{r_0}) \rangle_{\vec{r_0}}$$

$$C_{nm}(x, y) \rightarrow C_{nm}(r) := \langle u_n(\vec{r_0} - \vec{r}) \cdot u_m(\vec{r_0}) \rangle_{\vec{r_0}}$$

with $r = |\vec{r}|$

An example of such a radial symmetric height–height correlation function is shown in Fig. 4.2.

2.1.3.3 Membrane Displacement Correlation Functions

In analogy, a statistical, time-averaged description of the fluctuations of interacting membranes is provided by the membrane displacement correlation functions

$$g_{nm}(r) := \left\langle \left[u_n(\vec{r_0} - \vec{r}) - u_m(\vec{r_0}) \right]^2 \right\rangle_{\vec{r_0}}.$$

Lei et al. have derived an expression [54] which is valid in the special case of oriented multilayers that are infinitely expanded in all directions. However, this case does not apply to solid-supported membrane multilayers used for experiments. In Sect. 4.1 of this work a way to overcome this issue is presented.

2.1.4 Membrane Models

This thesis deals with several types of planar model systems of membranes and cell surfaces. Here, an introduction into selected oriented membrane models is given. Other commonly used, but non-oriented membrane models, such as unilamellar or multilamellar lipid vesicles are not discussed. Membrane models serve as simplified, well-defined models of biological membranes and allow for the detailed investigation of selected physical membrane properties. The choice of the membrane model determines which aspects of biological membranes can be studied.

2.1.4.1 Langmuir Lipid Monolayers

Langmuir lipid monolayers are insoluble monomolecular lipid films at the air/water interface [24, 25]. Here, as a consequence of the hydrophobic effect, the amphiphilic lipid molecules expose their hydrophobic tails to the gas phase.

This is illustrated in Fig. 2.6. The surface area available per lipid molecule, A, is controlled with a movable barrier.

By monitoring the surface pressure of the lipid film, π, as a function of A, the in-plane interaction of the molecules, confined in two dimensions, can be studied. Often, the lipid monolayer undergoes a phase transition as the A decreases, from a more disordered, liquid expanded (L_e, Fig. 2.6 top) phase to an ordered, liquid condensed (L_c, Fig. 2.6 bottom) phase, where the ordering refers to the hydrocarbon chains. This phase transition can be described with the two-dimensional analogue of the three-dimensional van-der-Waals (or Dieterici) equation, which accounts for the two phases and for a coexistence regime. An idealized Langmuir isotherm, i.e., a π versus A plot, recorded at a constant temperature, is schematically shown in Fig. 2.7. The curve shows the regimes of L_e and L_c phases, as well as a plateau region corresponding to the phase coexistence. As the surface area per molecule approaches to the minimum molecular are, denoted by A_0, the surface pressure diverges. The maximum achievable surface pressure is limited by the surface tension of the air/water interface ($\sigma \cong 72$ mN/m), but usually the monolayer collapses far below that value in practice. Often no phase transition is observed, due to the molecule-intrinsic incapability to form ordered structures of their hydrocarbon chains (e.g., double bonds in the hydrocarbon chains or bulky head groups), and the monolayer remains in L_e-phase throughout the isotherm.

When a lipid monolayer is compressed to a surface area per molecule, which is representative for the molecular area found in biological membranes, the hydrophilic monolayer surface exposed to the aqueous bulk phase constitutes a well-defined model of a membrane surface. This surface can be characterized by various (surface-sensitive) techniques (such as fluorescence microscopy, Brewster angle microscopy, X-ray and neutron scattering, Kelvin probe, ellipsometry, and interfacial shear rheometry) in terms of structural, dynamical, mechanical, and dielectric properties. The surface is accessible from the aqueous side and can be the starting point for studies on the interaction of various solutes (ions, proteins, drug molecules) with the membrane surface [26, 27].

Fig. 2.6 Sketch of a Langmuir film balance. In many cases, lipids undergo a transition from disordered L_e-phase to ordered L_c-phase upon compression

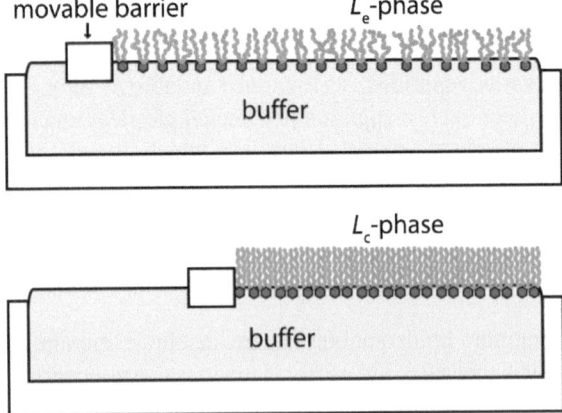

Fig. 2.7 Idealized Langmuir isotherm showing an L_e regime for large molecular areas, an L_c regime for small molecular areas, and a plateau region corresponding to the phase coexistence for intermediate molecular areas

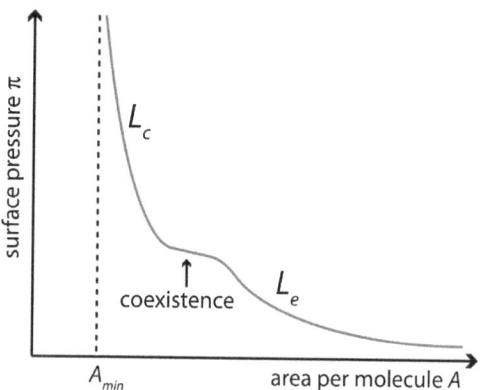

2.1.4.2 Solid-Supported Lipid Monolayers

Solid-supported lipid monolayers constitute models of biological membrane surfaces at the interface between a solid substrate and the aqueous phase [28]. This is illustrated in Fig. 2.8. Solid-supported lipid monolayers can be created by spreading small unilamellar lipid vesicles (SUVs) onto hydrophobic solid substrates under bulk buffer. The formation of the monolayer is driven by the hydrophobic effect. The resulting in-plane density of lipid molecules is very similar to that found in lipid bilayer membranes. Alternative methods for the deposition of lipid monolayers onto hydrophobic substrates are the Langmuir–Blodgett or Langmuir–Schaefer transfer [29], which enable the control of the in-plane density of the lipids. In contrast to Langmuir lipid monolayers, solid-supported lipid monolayers can be rotated in space, and therefore lend themselves towards scattering experiments, which often require sample rotations (see Sect. 3.3.1.1).

2.1.4.3 Solid-Supported Lipid Bilayers

Solid-supported lipid bilayers constitute well-defined models of biological lipid membranes confined in two dimensions [30]. This is illustrated in Fig. 2.9 (left).

Fig. 2.8 Sketch of solid-supported monolayer deposited onto a flat hydrophobic solid substrate

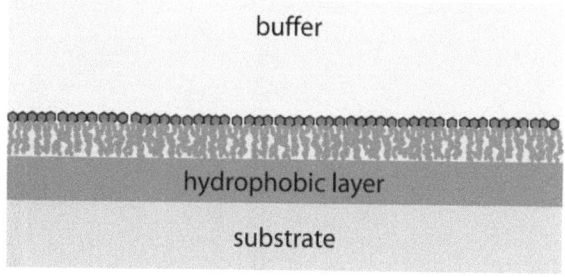

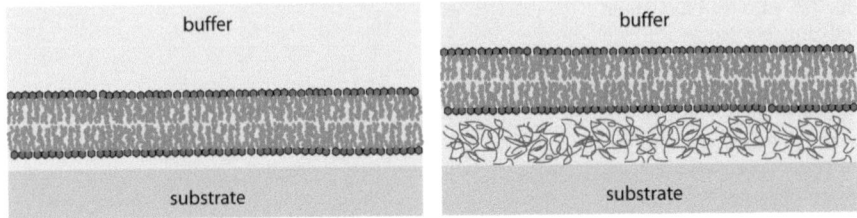

Fig. 2.9 Sketch of a solid-supported (*left*) and a polymer-supported (*right*) lipid bilayer

They are well suited to study the diffusion of lipids and membrane-associated proteins [31]. Solid-supported lipid bilayers can be created by the fusion of SUVs, or by Langmuir–Blodgett/Langmuir–Schaefer transfers. The bilayer is confined at the solid surface due to strong van der Waals attraction. To reduce the influence of this strong interaction with the substrate on the bilayer behavior, and to create a more physiological environment for the bilayer, soft polymer interlayers between solid support and bilayer (see Fig. 2.9 right) have gained importance in recent years [32, 33].

2.1.4.4 Solid-Supported Membrane Multilayers

Solid-supported membrane multilayers constitute model systems of interacting biological membranes. This is illustrated in Fig. 2.10. These systems, which can be prepared either by evaporating organic lipid solutions on the solid surface or by multiple Langmuir–Blodgett transfers, can be studied in humidified air but also under bulk buffer solution, if they retain their oriented lamellar structure under these conditions. This is the case if a finite equilibrium membrane separation is established by the inter-membrane interactions (see Sect. 2.1.2). To investigate structure and mechanics of interacting lipid membranes, the samples can be studied by X-ray and neutron scattering techniques (see Chaps. 5, 6).

Fig. 2.10 Sketch of solid-supported multilayers aligned parallel with the flat substrate

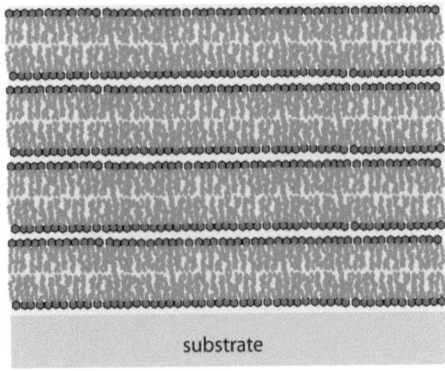

2.2 Principles of X-Ray and Neutron Scattering

X-ray and neutron scattering techniques are the principal experimental methods applied in this thesis. In the following, the common theoretical concepts of the used scattering techniques are introduced. Since this thesis deals with oriented membrane models, the main focus is put on specular and off-specular scattering from planar, oriented systems.

2.2.1 Basic Principles

X-ray and neutron scattering techniques are ideal for the structural characterization of a sample at molecular length scales. They provide structural information with high spatial resolution in a non-destructive manner and have access to buried structures in contrast to other techniques. A good introduction to X-ray and neutron scattering is given in the following references [34–37]. When probing a sample at length scales which are large compared to atomic structures (≈ 1 Å), the description of the sample in terms of continuous media has proven very powerful. Here, the atomic structure of the sample is neglected and the sample is parameterized with a refractive index $n(\vec{r})$, which is a complex function of the spatial coordinates $\vec{r}$:

$$n = 1 - \delta + i\beta$$

$$\delta = \frac{\lambda^2}{2\pi}\rho$$

$$\beta = \frac{\lambda}{4\pi}\mu$$

λ denotes the wavelength of the X-ray or neutron beam, ρ the scattering length density (SLD) of the medium, and μ the absorption coefficient of the medium for the beam. In case of X-rays, μ mainly accounts for the photoelectric consumption of X-ray photons, while the SLD is proportional to the electron density $\rho = r_0\rho_{el}$, where r_0 denotes the classical electron radius (or Thomson scattering length). In case of cold or thermal neutrons, μ accounts for neutron capture processes, while the SLD depends on the nuclear composition of the medium:

$$\rho = \sum_j \rho_j b_j$$

Here, ρ_j denotes the volume density of the nuclide species j, and b_j its coherent scattering length, which is tabulated for the vast majority of nuclides [38].

2.2.1.1 X-Ray versus Neutrons

Commonly used X-ray and neutron beams have comparable wavelengths in the Armstrong range. However, the ways X-rays and neutrons interact with matter are fundamentally different. While X-rays are scattered by the electron shells of atoms and molecules via electromagnetic interactions, neutrons are scattered by the atomic nuclei via strong interaction and the interaction between the magnetic dipole moments of the nucleus and the neutron [39]. Fortunately, despite these differences, the essentially same mathematical formalism can be used for both radiation types if the scattering signals are interpreted using the before mentioned continuum description. Besides these fundamental differences there are quite a few practical differences which become important in an experiment. Currently, the photon flux provided by modern X-ray sources (e.g., synchrotrons) is by orders of magnitude higher than the neutron flux provided by the most powerful neutron sources (nuclear reactors and spallation sources). For this reason, to record scattering signals with a desired signal to noise ratio typically takes much longer with neutrons, and, as a consequence, the spatial resolution achieved with X-rays is generally higher. On the other hand, the absorption coefficients of the matter are typically much lower for neutrons than for X-rays. This renders neutrons superior for the investigation of deeply buried structures. Moreover, the fact that the neutron scattering length can strongly differ between two isotopes of the same chemical element offers the unique possibility to manipulate the scattering contrast within the sample. This method is known as "contrast variation", and is particularly powerful for soft-matter samples and biological samples. Here, molecules or parts of molecules can be highlighted by isotopic labeling without significantly influencing their chemical behavior. This works particularly well when hydrogen is replaced by deuterium (called deuteration), as these nuclides possess extremely different scattering lengths. Generally, whether X-rays or neutrons are more suited for an experiment depends on the sample and on the experimental conditions, and in many cases it is desirable to investigate a sample with both methods.

2.2.1.2 Scattering from Oriented Planar Samples

Oriented planar samples (e.g., thin films on a flat substrate) offer the possibility to distinguish between out-of-plane (z) and in-plane (x and y) directions. Here, the scattering signals contain structural information simultaneously for both out-of-plane and in-plane directions. The former can be extracted from the specular scattering intensity, while the latter is contained in the off-specular (diffuse) scattering intensity [34, 40–43].

Oriented samples are commonly modeled with "slab models" [44–47], where each slab represents a layer of constant refractive index n. Such a slab model is illustrated in Fig. 2.11. Since, for X-rays and thermal neutrons all materials possess refractive indices close to unity and, in thin layers, absorption can be neglected, the sample structure is typically described in terms of SLD profiles

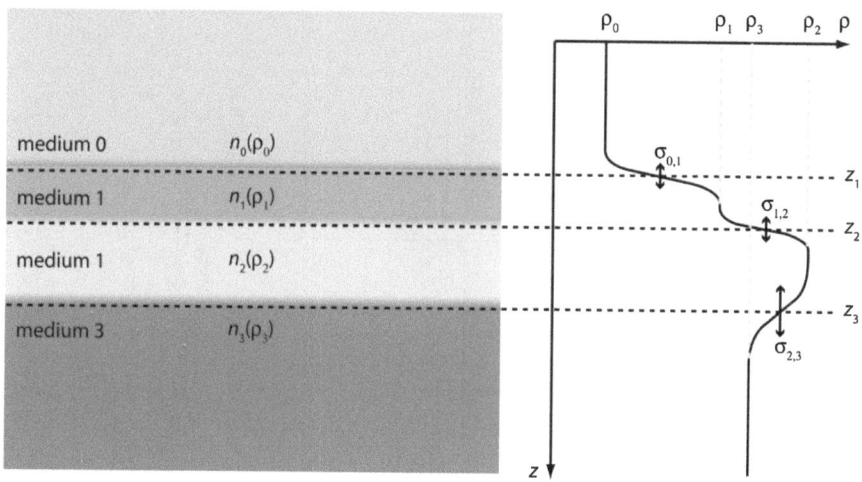

Fig. 2.11 (*Left*) schematic slab model illustration of an oriented sample. (*Right*) corresponding SLD profile, parameterized by the constant SLDs ρ of the stratified media and the SLD gradients across the interfaces (characterized by the width σ)

rather than in terms of refractive index profiles. The continuous transition of ρ at the interface between a slab and its neighbors is typically accounted for with an error function (characterized by transition width σ). Figure 2.12 shows the geometry considered in the following for a scattering experiment with an oriented sample.

Let a monochromatic beam with wave vector $\vec{k_i}$ impinge to an oriented sample with the incident angle θ_i. The plane of incidence is defined by $\vec{k_i}$ and the normal to the sample. We further consider that the beam is scattered into a direction which is part of the plane of incidence[5] and has an angle θ_f with the sample surface as indicated in Fig. 2.12. $\vec{k_f}$ denotes the wave vector of the scattered beam. In the here considered elastic scattering processes the length of the wave vector is conserved:

$$\left|\vec{k_f}\right| = \left|\vec{k_i}\right| = \frac{2\pi}{\lambda}$$

The momentum transfer corresponding to the scattering process is characterized by the scattering vector $\vec{q}$:

$$\vec{q} = \vec{k_f} - \vec{k_i},$$

---

[5] Within the frame of this thesis and for the explanations below it is not necessary to consider the scattering into directions which are not part of the plane of incidence.

Fig. 2.12 Sketch of the scattering geometry considered in this thesis

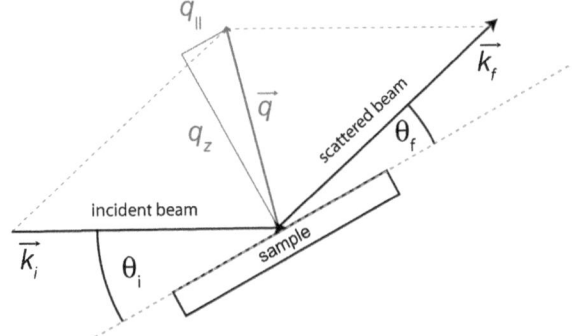

which can be decomposed into an out-of-plane component q_z, and a in-plane component $q_{||}$. These two components are the reciprocal space coordinates and can be expressed in terms of the scattering angles θ_i and θ_f:

$$q_z = k_f^z - k_i^z = \frac{2\pi}{\lambda}(\sin\theta_f + \sin\theta_i),$$

$$q_{||} = k_f^{||} - k_i^{||} = \frac{2\pi}{\lambda}(\cos\theta_f - \cos\theta_i)$$

2.2.2 Specular Scattering

The (in-plane averaged) out-of-plane structure of an oriented sample can be extracted from the specular scattering intensity, where only the intensity of the mirror-like reflection is considered (i.e., $\theta_f = \theta_i =: \theta$). In this case $q_{||} = 0$ and q_z simplifies to:

$$q_z = \frac{4\pi}{\lambda}\sin\theta$$

2.2.2.1 Specular Reflectivity from a Single Ideal Interface

We consider an ideal (i.e., perfectly smooth) planar interface between two homogenous media denoted with 0 and 1, and a beam impinging onto the interface coming from medium 0. The out-of-plane component k_j^z of the wave vector in each medium depends on q_z and on the refractive indices of the media, n_0 and n_1:

$$k_j^z = \frac{2\pi}{\lambda}\sqrt{\left(\frac{q_z\lambda}{4\pi}\right)^2 + n_j^2 - 1}$$

This can be simplified if absorption is neglected:

$$k_j^z = \sqrt{\left(\frac{q_z}{2}\right)^2 - 4\pi\rho_j}$$

The complex amplitudes of the reflected and transmitted waves are given as the Fresnel amplitude reflection and transmission coefficient:

$$r_{0,1}^F = \frac{k_0^z - k_1^z}{k_0^z + k_1^z} \quad \text{and} \quad t_{0,1}^F = 1 + r_{0,1}^F.$$

In contrast to the case of visible light, these coefficients do not depend significantly on the polarization of the X-ray or neutron beam [34]. The specular reflectivity R, defined as the ratio between the reflected intensity I_r and the incident intensity I_i, is the absolute square of the complex Fresnel amplitude reflection coefficient:

$$R(q_z) := I_f/I_i = \left| r_{0,1}^F \right|^2$$

If the refractive index of medium 1 is lower than that of medium 0, the beam is totally reflected ($R = 1$) below a critical q_z value, denoted with q_z^c. This value can be approximated in terms of the SLDs of the media:

$$q_z^c \cong \sqrt{16\pi(\rho_1 - \rho_0)}$$

Below q_z^c, medium 1 is illuminated only in close vicinity of the interface is by an evanescent field of q_z-dependent intensity decay length Λ.

$$\Lambda = \Lambda_0 \left(1 - \left(q_z/q_z^c \right)^2 \right)^{-1/2}, \quad \text{with } \Lambda_0 = 1/q_z^c$$

This dependency is shown in Fig. 2.13, and it is seen that the decay length diverges as q_z approaches q_z^c. Above q_z^c the beam propagates into medium 1.

In Fig. 2.14 (left) the reflectivity of an ideal interface ($\rho_0 = 0$, $\rho_1 = 20 \times 10^{-6}$ Å^{-2}, $n_0 > n_1$) is plotted as a function of q_z. For high q_z values ($q_z \gg q_z^c$), the reflected intensity is weak. In this limit the reflectivity decays as q_z^{-4}, see dashed

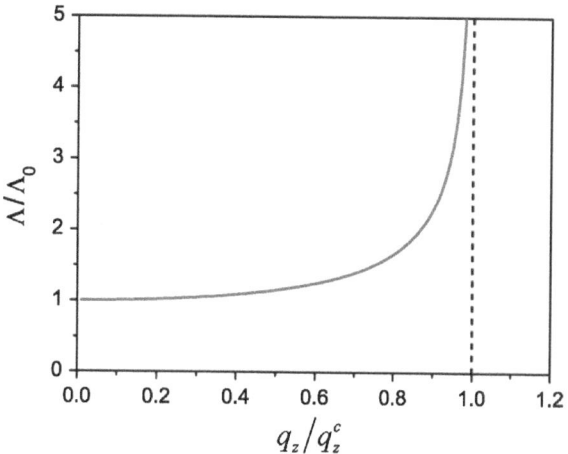

Fig. 2.13 Intensity decay length of the evanescent field as a function of q_z. The decay length diverges as q_z approaches q_z^c

line in Fig. 2.14 (left). In the regime of weak reflection the kinematic approximation, also known as First Born Approximation is valid. In this approximation, the incident beam is assumed to be "inexhaustible" and the reduction of the transmitted intensity due to reflections is neglected. Within the range of its validity, the kinematic approximation is very powerful, as it enables the treatment of many scattering problems in terms of simple algebraic expressions. However, the so calculated scattering intensities cannot be scaled with respect to the incident beam intensity.

2.2.2.2 Interfacial Roughness

Realistic (non-ideal) interfaces between two homogenous media do not possess a sharp jump in the SLD (and thus, in the refractive index) but a gradual transition (see Fig. 2.11). To describe this transition, error functions characterized by the transition width σ are typically used. For this case, Nevot and Croce [48] have derived how the amplitude reflection coefficient for an interface should be modified:

$$r_{0,1}^{F} \rightarrow r_{0,1}^{F} \cdot \exp\left(-\frac{1}{2}q_z^2\sigma_{0,1}^2\right)$$

The effect of the interfacial roughness is shown in Fig. 2.14 (right) for $\sigma = 3$ Å. It should be pointed out that the specular intensity contains information only on the in-plane averaged sample structure. This averaged structure does not distinguish between an SLD gradient across the interface (i.e., an identical z-dependence of the SLD for all x and y) and topological roughness (i.e., the in-plane heterogeneity of the z-position of the interface). Only the interpretation of off-specular (diffuse) scattering intensities enables this distinction (see Sect. 2.2.3).

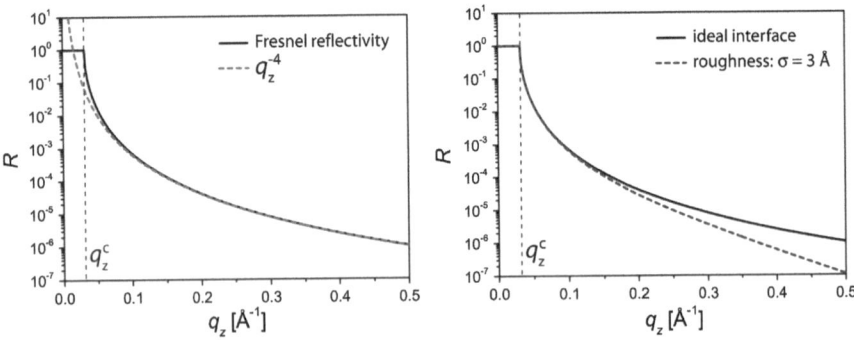

Fig. 2.14 (*Left*) Fresnel reflectivity from a single ideal interface. (*Right*) influence of interfacial roughness

2.2.2.3 Stratified Interfaces

The reflectivity signals from N stratified interfaces (see Fig. 2.11) can be calculated using the formalisms introduced by Parratt [34, 49]. Based on the Fresnel amplitude reflection coefficients,

$$r^F_{j,j+1} = \frac{k^z_j - k^z_{j+1}}{k^z_j + k^z_{j+1}} \cdot \exp\left(-\frac{1}{2}q^2_z \sigma^2_{j,j+1}\right)$$

the effective amplitude reflection coefficient of the stratified system $r_{0,N}$ is calculated recursively starting from the last interface (between medium $N-1$ and medium N):

$$r_{N-1,N} = r^F_{N-1,N}$$

$$r_{j,N} = \frac{r^F_{j,j+1} + r_{j+1,N} \cdot e^{2ik^z_{j+1}d_{j+1}}}{1 + r^F_{j,j+1} \cdot r_{j+1,N} \cdot e^{2ik^z_{j+1}d_{j+1}}}$$

j counting downwards from $N-2$ to 0, $d_j = z_{j+1} - z_j$ denotes the thickness of medium j.

The reflectivity is then given as: $R(q_z) = |r_{0,N}|^2$.

A calculated reflectivity signal from 3 stratified interfaces ($\rho_0 = 0$, $\rho_1 = 10 \times 10^{-6}$ Å^{-2}, $\rho_2 = 25 \times 10^{-6}$ Å^{-2}, $\rho_2 = 20 \times 10^{-6}$ Å^{-2}, $d_1 = 35$ Å, $d_2 = 25$ Å, $\sigma_{0,1} = \sigma_{1,2} = \sigma_{2,3} = 3$ Å) is presented in Fig. 2.15. In contrast to the signal from a single interface, the shown curve possesses various features like oscillations and deep minima (known as Kiessig fringes), caused by the interference of the waves reflected at different interfaces. The Parratt formalism constitutes a fully dynamical description of the scattering process where the intensity of the incident beam is conserved, and the reflections from each interface (including multiple back- and forth reflections within the slabs) as well as the interference of

Fig. 2.15 Specular reflectivity from a set stratified rough interfaces

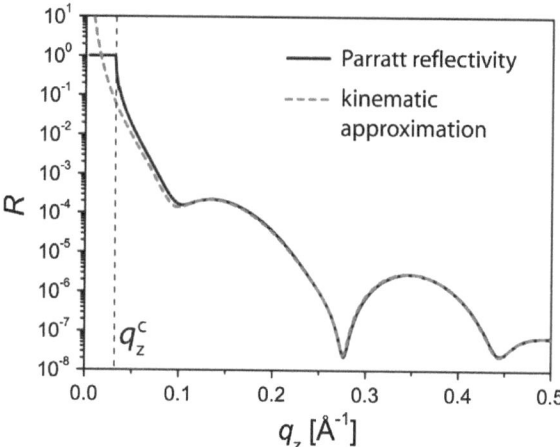

all reflected and transmitted waves are considered. An equivalent formalism was presented by Abeles, based on layer matrices [50]. In the regime of weak reflection ($q_z \gg q_z^c$), the reflectivity signals from stratified interfaces can be expressed in the kinematical approximation using the Master Formula [34]. Here, the signal is described with the asymptotic scaling of the reflectivity of a single, ideal interface, and with the Fourier-transformed SLD gradient.[6]

$$R(q_z) \propto \frac{1}{q_z^4} \left| \int_{-\infty}^{\infty} \frac{d\rho}{dz} \exp(-iq_z z) dz \right|^2 \tag{2.1}$$

In case of slab models, this expression can be calculated very efficiently, as the derivatives of error functions are Gaussian functions that can be Fourier-transformed analytically. The resulting signal is also shown in Fig. 2.15. As the signal possesses no absolute scale, it was multiplied by a pre-factor such that it coincides with the dynamically calculated curve in the valid range.

2.2.2.4 Periodic Multilayers

In case many interfaces are arranged periodically ($z_{j+1} - z_j = d$ for all j), the interference of the waves reflected from the interfaces results in peaks of high reflection intensity, known as Bragg peaks. The positions of the peaks maxima, q_z^{max}, are located periodically along q_z:

$$q_z^{max} = 2m\pi/d, \quad m = 1, 2, \ldots \tag{2.2}$$

This is shown in Fig. 2.16 (solid black line) for a periodic system of 100 double layers ($\rho_A = 20 \times 10^{-6}\,\text{Å}^{-2}$, $\rho_B = 0$, $d_A = 35\,\text{Å}$, $d_B = 25\,\text{Å}$) between two semi-infinite media ($\rho_0 = 0$, $\rho_1 = 20 \times 10^{-6}\,\text{Å}^{-2}$) and $\sigma = 3\,\text{Å}$ for all interfaces. The intensity of the Bragg peaks (especially of the first Bragg peak with $m = 1$) often becomes comparable to that of the incident beam. In this case the kinematic approximation fails even for $q_z \gg q_z^c$, as it predicts too high intensities (see dashed red line in Fig. 2.16).

2.2.3 Off-Specular (Diffuse) Scattering

Real interfaces always possess topological roughness, which leads non-mirror-like scattering ($\theta_f \neq \theta_i$, $q_\parallel \neq 0$, see Fig. 2.12), called off-specular or diffuse scattering. To consider diffuse scattering, the intensity has to be described as a function of q_z and $q_\parallel$. Like the reflectivity $R(q_z)$, the (more general) scattering intensity

---

[6] Often the asymptotic q_z^{-4} factor is replaced by the reflectivity curve of an ideal interface between the semi-infinite media. The results of this "hybrid" description come close to those of the full dynamical treatment in some cases.

Fig. 2.16 Specular reflectivity from a multilayered system. The signal exhibits Bragg maxima periodic in q_z. The breakdown of the kinematic approximation at the first ($m = 1$) Bragg peak is highlighted with a *blue circle*

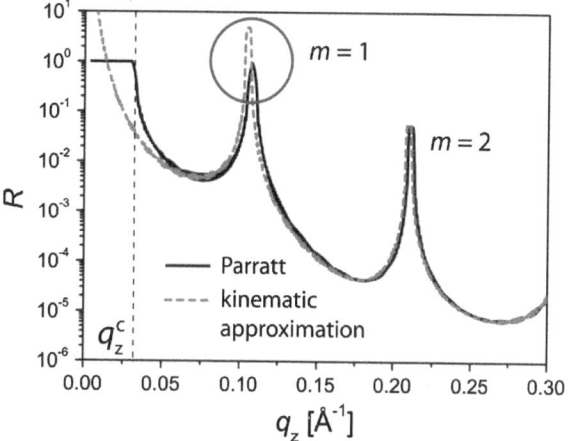

$I(q_z, q_\parallel)$ is defined in terms of total intensity in the following. It can be expressed with the scattering function $S(q_z, q_\parallel)$, which is defined in terms of incident flux and the solid angle of detection [43]:

$$I(q_z, q_\parallel) \propto \frac{1}{q_z^2} S(q_z, q_\parallel)$$

2.2.3.1 Scattering from a Single, Topologically Rough Interface

The topological roughness of an interface can be parameterized with the local out-of-plane displacement $u(x, y)$: $= z(x, y) - \langle z \rangle$, were $z(x, y)$ denotes the height of the interface at the point (x, y) and $\langle z \rangle$ its average height. The diffuse scattering from an interface is determined by the characteristics of the topological roughness. The scattering function can be expressed [34, 40, 42, 43] in terms of the height–height correlation $C(r)$ of the interface (see Sect. 2.1.3.2). In Kinematical approximation, the scattering function is given as:

$$S(q_z, q_\parallel) \propto \frac{e^{-q_z^2 \sigma^2}}{q_z^2} \int_{-\infty}^{\infty} e^{q_z^2 C(r)} e^{-i q_\parallel r} \, dr$$

The expression comprises a Fourier-transformation which has to be evaluated numerically in general. Figure 2.17 (left) shows the calculated total (specular and diffuse) scattering intensity $I(q_z, q_\parallel)$ from a single topologically rough interface ($\rho_0 = 0$, $\rho_1 = 20 \times 10^{-6}$ Å^{-2}) in color coded logarithmic scale. Here, the height–height correlation function was parameterized[7] as $C(r) = \sigma^2 \exp\left(-(r/\xi)^{2h}\right)$, with

---

[7] Such a parameterization is commonly used to describe interfaces with self-affine roughness [40], but cannot describe all types of surface topologies (see Sect. 4.1.1).

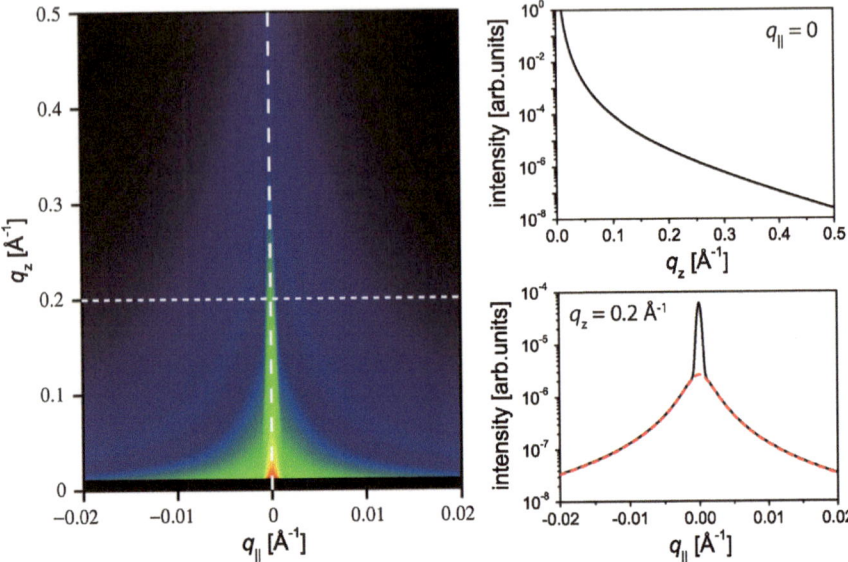

Fig. 2.17 (*Left*) simulated scattering intensity from a single rough interface in kinematic approximation. (*Top right*) central cut through the intensity map along the specular line (*vertical broken line in the left panel*). (*Bottom right*)(*solid black line*) intensity as a function of $q_{\|}$ (*horizontal dashed line in the left panel*). (*Dashed red line*) diffuse intensity alone

the topological rms roughness $\sigma = 3$ Å, the decay length $\xi = 500$ Å, and the stretching exponent $h = 0.5$. The top right panel shows the intensity along the specular line as a function of q_z. The bottom right panel of the figure (solid black line) shows a cut through intensity map along $q_{\|}$ at $q_z = 0.2$ Å^{-1}, indicated by the horizontal dashed line in the left panel. The central maximum represents the specular peak, where the intensity is dominated by the specular contribution. Except for the peak region, the intensity corresponds to the diffuse contribution. For practical reasons, the width of the specular peak (ideally a Dirac delta function of $q_{\|}$) was broadened by convoluting $I(q_z, q_{\|})$ with an artificial point-spread function in $q_{\|}$. The dashed red line represents the diffuse intensity alone.

A dynamical treatment of the scattering from a topologically rough interface requires taking refraction effects and strong scattering into account. This is achieved by the Distorted Wave Born Approximation (DWBA) [40, 42], at the cost of a much higher numerical effort. In DWBA, the scattering function is given as:

$$S\left(q_z, q_{\|}\right) \propto \left|t_{0,1}^{F,i}\right|^2 \left|t_{0,1}^{F,f}\right|^2 \frac{1}{q_z^2} \exp\left(-\frac{1}{2}\sigma^2\left((q_1)^2 + (q_1^*)^2\right)\right) \cdot \int\limits_{-\infty}^{\infty} e^{|q_1|^2 C(r)} e^{-iq_{\|}r} dr$$

$t_{0,1}^{F,i}$ and $t_{0,1}^{F,f}$ denote the Fresnel amplitude transmission coefficients for the incident and scattered beam, respectively:

$$t_{0,1}^{F,i} = 1 + \frac{k_{i,0}^z - k_{i,1}^z}{k_{i,0}^z + k_{i,1}^z}, \, t_{0,1}^{F,f} = 1 + \frac{k_{f,0}^z - k_{f,1}^z}{k_{f,0}^z + k_{f,1}^z}$$

$q_1 = k_{f,1}^z - k_{i,1}^z$ denotes the momentum transfer in medium 1.

These quantities are defined via the out-of-plane components of the incident and scattered wave vectors in medium 1, $k_{i,1}^z$ and $k_{f,1}^z$, which can be calculated by mathematically inverting the scattering geometry:

$$k_{i,1}^z = \sqrt{(k_i^z)^2 - 4\pi\rho_1}, \quad k_{f,1}^z = \sqrt{\left(k_f^z\right)^2 - 4\pi\rho_1}$$

$$k_i^z = k_0 \sin\theta_i, \quad k_f^z = k_0 \sin\theta_f$$

$$\theta_i = \frac{1}{2}\arccos\left(1 - \frac{q_x^2 + q_z^2}{2k_0^2}\right) + \arctan\left(\frac{q_x}{q_z}\right), \theta_f$$

$$= \frac{1}{2}\arccos\left(1 - \frac{q_x^2 + q_z^2}{2k_0^2}\right) - \arctan\left(\frac{q_x}{q_z}\right)$$

$$k_0 = 2\pi/\lambda$$

Figure 2.18 (left) shows the total scattering intensity $I(q_z, q_\parallel)$ from a single topologically rough interface (same parameters as before) calculated in DWBA.

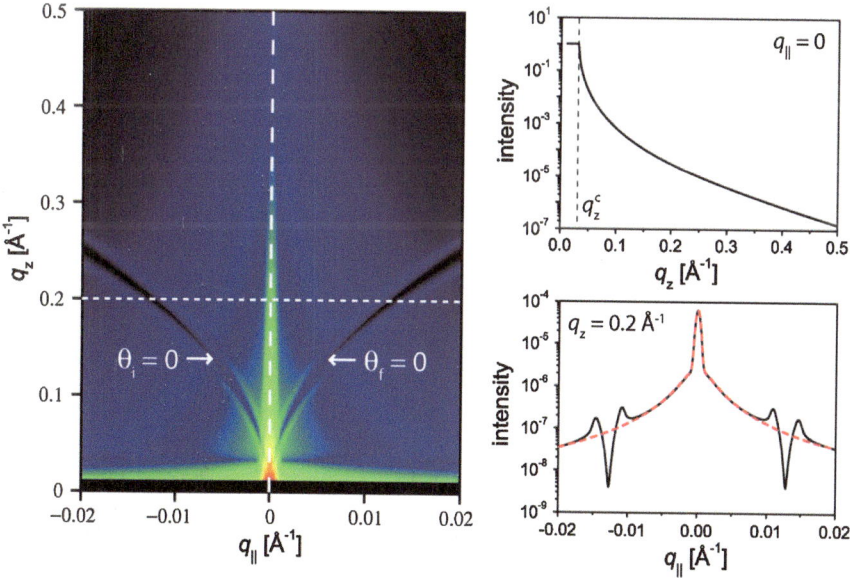

Fig. 2.18 (*Left*) simulated scattering intensity from a single rough interface in DWBA. (*Top right*) Central cut through the intensity map along the specular line (*vertical broken line in the left panel*). (*Bottom right solid black line*) intensity as a function of $q_\parallel$ (*horizontal dashed line in the left panel*). *Dashed red line*: kinematic approximation

The top right panel shows the intensity along the specular line as a function of q_z, which exhibits the total reflection regime in the dynamic treatment. The reciprocal space map contains additional features, most prominently the Yoneda-wings along the sample horizons ($\theta_i = 0$, $\theta_f = 0$) as indicated in the figure. They are results of the refraction effects which become important at grazing incidences near the sample horizons. Again, the right panel of the figure (solid black line) shows a cut through intensity map along $q_\parallel$ at $q_z = 0.2$ Å^{-1}, indicated by the horizontal dashed line in the left panel. In this representation, the Yoneda wings are seen as dips in the intensity. For comparison, the result of the kinematic approximation is superimposed to the graph (dashed red line).

2.2.3.2 Scattering from Stratified Interfaces with Correlated Topological Roughness

In kinematic approximation, the scattering function of a set of N stratified interfaces with correlated topological roughness is given as:

$$S(q_z, q_\parallel) \propto \frac{1}{q_z^2} \sum_{n=1}^{N} \sum_{m=1}^{N} e^{-\frac{1}{2}q_z^2\left(\sigma_n^2 + \sigma_m^2\right)} \Delta\rho_n \Delta\rho_m e^{-iq_z(z_n - z_m)} \int_{-\infty}^{\infty} e^{q_z^2 C_{nm}(r)} e^{-iq_\parallel r} \, dr$$

with $C_{nm}(r) = \left\langle u_n\left(\vec{r_0} - \vec{r}\right) \cdot u_m\left(\vec{r_0}\right)\right\rangle_{\vec{r_0}}$ and $r = |\vec{r}|$ (see Sect. 2.1.3)

$\Delta\rho_n$ denotes the SLD jump across the nth interface and σ_n its topological rms roughness. In the following, the slab model introduced for the calculation of the specular reflectivity is used (see Sect. 2.2.2.3). Further, we assume that each interface possesses the height–height self-correlation $C_{n=m}(r) = \sigma^2 \exp\left(-(r/\xi)^{2h}\right)$ with the identical parameters $\sigma = 3$ Å, $\xi = 500$ Å, and $h = 0.5$ (like the single topologically rough interface discussed before). To highlight the role of the cross-correlation terms $C_{n \neq m}(r)$, two extreme scenarios are considered: (1) Full correlation between all interfaces. The calculated scattering intensity $I(q_z, q_\parallel)$ for this scenario is shown in the top left panel of Fig. 2.19. (2) Two interfaces possess fully correlated roughness, while there is no correlation between these two interfaces and the third one. The corresponding scattering intensity is shown in the top right panel of Fig. 2.19. Mathematically this can be expressed using the correlation matrices:

$$C_{nm} = C_{n=m} \begin{pmatrix} 1 & 1 & 1 \\ 1 & 1 & 1 \\ 1 & 1 & 1 \end{pmatrix} \tag{1}$$

$$C_{nm} = C_{n=m} \begin{pmatrix} 1 & 0 & 0 \\ 0 & 1 & 1 \\ 0 & 1 & 1 \end{pmatrix} \tag{2}$$

As can be seen in the figure, there are distinct differences in the diffuse scattering intensity. These differences can be highlighted by plotting the scattering intensity as a function of q_z for various $q_{\parallel}$ values. This is shown in the bottom panels of Fig. 2.19, where the scattering intensity along q_z is plotted for $q_{\parallel} \cong 0$ (central cut, see dashed white lines in top panels) and $q_{\parallel} \neq 0$ (diffuse cut, see broken white lines in top panels). In the fully correlated case 1 (bottom left panel), all features (minima and maxima) seen in the central cut (where the intensity is dominated by the specular contribution) are also reflected in the diffuse cut, which is constituted by the diffuse intensity. In contrast, in the partially correlated case 2 (bottom right panel), only the minimum at around $q_z = 0.27$ Å^{-1} of the central cut

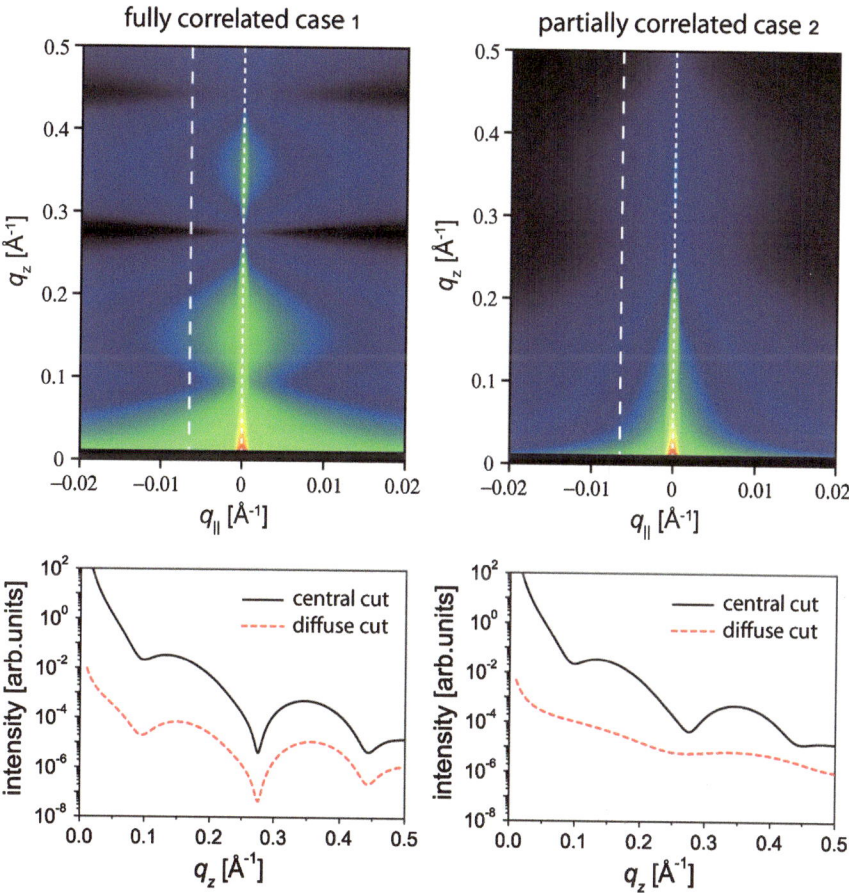

Fig. 2.19 (*Top*) simulated scattering intensity from a set of interfaces with correlated topological roughness in kinematic approximation. The *vertical dashed lines* indicate the "central cut", while the *vertical broken lines* indicate the "diffuse cut". (*bottom*) "Central cuts" and "diffuse cuts" through the intensity maps. (*left*) Case 1 with full correlation between all interfaces. (*right*) Case 2 with only partial correlation

is also seen in the diffuse cut, and this minimum corresponds to the interference between the two correlated interfaces. Generally, the way how the features of the diffuse scattering intensity relate to those of the specular scattering intensity is determined by the cross-correlation functions.

The dynamical treatment of the scattering from stratified interfaces with correlated topological roughness requires the use of the DWBA. This enables the correct description of strong scattering intensities and features which are characteristic for multiple scattering. However, the numerical effort for the computations is enormous and the corresponding equations are not presented in this work. For further reading the following literature can be consulted [40, 42, 51, 52].

References

1. J.N. Israelachvili, *Intermolecular and Surface Forces* (Academic Press Inc., London, 1991)
2. J.N. Israelachvili, D.J. Mitchell, B.W. Ninham, Theory of self-assembly of hydrocarbon amphiphiles into micelles and bilayers. J. Chem. Soc. Faraday Trans. II **72**, 1525 (1976)
3. E. Sackmann, in *Structure and Dynamics of Membranes*, ed. by R. Lipowski, E. Sackmann (Elsevier, Amsterdam, 1995)
4. R. Koynova, M. Caffrey, Phases and phase transitions of the phosphatidylcholines. Biochim. Biophys. Acta **1376**, 91 (1998)
5. D. Marsh, General features of phospholipid phase transitions. Chem. Phys. Lipids **57**, 109 (1991)
6. A. Ben-Shaul, in *Structure and Dynamics of Membranes*, ed. by R. Lipowski, E. Sackmann (Elsevier, Amsterdam, 1995)
7. B.V. Derjaguin, N.V. Churaev, *Surface Forces* (Consultants Bureau, New York, 1987)
8. B.V. Derjaguin, L.D. Landau, A theory of the stability of strongly charged lyophobic sols and the coalescence of strongly charged particles in electrolytic solution. Acta Physicochim. URSS **14**, 633 (1941)
9. P.C. Hiemenz, *Principles of Colloid and Surface Chemistry* (Dekker, New York and Basel, 1977)
10. E.J. Verwey, J.T.G. Overbeek, *Theory of Stability of Lyophobic Colloids* (Elsevier, Amsterdam, 1948)
11. R.P. Rand, V.A. Parsegian, Hydration Forces between Phospholipid-Bilayers. Biochim. Biophys. Act. **988**, 351 (1989)
12. I.E. Dzyaloshinskii, E.M. Lifshitz, L.P. Pitaevskii, General theory of van der Waals' forces. Adv. Phys. **10**, 165 (1961)
13. E.M. Lifshitz, The theory of molecular attractive forces between solids. Soviet Phys. JETP (Engl. Transl.) **2**, 73 (1956)
14. E. Evans, D. Needham, in *Physics of Amphiphilic Layers*, vol. 21, ed. by J. Meunier, D. Langevin (Springer, Berlin, 1987)
15. B.W. Ninham, V.A. Parsegian, Van der Waals forces: special characteristics in lipid–water systems and a general method of calculation based on the Lifshitz theory. Biophys. J. **10**, 646 (1970)
16. B. Demé, M. Dubois, T. Zemb, Swelling of a lecithin lamellar phase induced by small carbohydrate solutes. Biophys. J. **82**, 215 (2002)
17. D.M. LeNeveu, R.P. Rand, V.A. Parsegian, Measurements of forces between lecithin bilayers. Nature **259**, 601 (1976)
18. L.J. Lis, M. McAlister, N. Fuller, R.P. Rand, V.A. Parsegian, Interactions between neutral phospholipid bilayer membranes. Biophys. J. **37**, 657 (1982)

19. M. Bachmann, H. Kleinert, A. Pelster, Fluctuation pressure of a stack of membranes. Phys. Rev. E **63**, 051709 (2001)
20. W. Helfrich, Steric interaction of fluid membranes in multilayer systems. Z. Naturforsch. **33a**, 305 (1978)
21. W. Helfrich, Lyotropic lamellar phases. J. Phys. Condens. Matter **6**, A79 (1994)
22. V.A. Parsegian, N. Fuller, R.P. Rand, Measured work of deformation and repulsion of lecithin bilayers. Proc. Natl. Acad. Sci. USA **76**, 2750 (1979)
23. L.D. Landau, E.M. Lifshitz, *Statistische Physik Teil 1* (Akademie Verlag, Berlin, 1987)
24. H. Möhwald, in *Structure and Dynamics of Membranes*, ed. by R. Lipowski, E. Sackmann (Elsevier, Amsterdam, 1995)
25. F. Rehfeldt, R. Steitz, S.P. Armes, R.v. Klitzing, A.P. Gast, M. Tanaka, Reversible activation of Diblock copolymer monolayers at the interface by pH modulation, 1: lateral chain density and conformation. J. Phys. Chem. B **110**, 9171 (2006)
26. R.G. Oliveira et al., Physical mechanisms of bacterial survival revealed by combined grazing-incidence X-ray scattering and Monte Carlo simulation. C. R. Chim. **12**, 209 (2009)
27. M.F. Schneider, K. Lim, G.G. Fuller, M. Tanaka, Rheology of glycocalix model at air/water interface. Phys. Chem. Chem. Phys. **4**, 1949 (2002)
28. D. Gassull, A. Ulman, M. Grunze, M. Tanaka, Electrochemical sensing of membrane potential and enzyme function using gallium arsenide electrodes functionalized with supported membranes. J. Phys. Chem. B **112**, 5736 (2008)
29. I. Langmuir, V.J. Schaefer, Activities of urease and pepsin monolayers. J. Am. Chem. Soc. **60**, 1351 (1938)
30. M. Tanaka, E. Sackmann, Polymer-supported membranes as models of the cell surface. Nature **437**, 656 (2005)
31. T. Schubert, M. Bärmann, M. Rusp, W. Gränzer, M. Tanaka, Diffusion of glycosylphosphatidylinositol (GPI)-anchored bovine prion protein (PrPc) in supported lipid membranes studied by single-molecule and complementary ensemble methods. J. Membr. Sci. **231**, 61 (2008)
32. M. Tanaka, S. Kaufmann, J. Nissen, M. Hochrein, Orientation selective immobilization of human erythrocyte membranes on ultrathin cellulose films. Phys. Chem. Chem. Phys. **3**, 4091 (2001)
33. M. Tanaka, E. Sackmann, Supported membranes as biofunctional interfaces and smart biosensor platforms. Phys. Stat. Sol. **203**, 3452 (2006)
34. J. Als-Nielsen, D. McMorrow, *Elements of modern X-ray physics* (Wiley, Chichester, 2001)
35. A.-J. Dianoux, G. Lander, *Neutron Data Booklet* (Old City Publishing, Philadelphia, 2003)
36. T.P. Russell, X-ray and neutron reflectivity for the investigation of polymers. Mater. Sci. Rep. **5**, 171 (1990)
37. M. Tolan, *X-Ray Scattering from Soft-Matter Thin Films: Materials Science and Basic Research* (Springer, New York, 1999)
38. V.F. Sears, Neutron scattering lengths and cross sections. Neutron News **3**, 26 (1992)
39. B. Povh, K. Rith, C. Scholz, F. Zetsche, *Teilchen und Kerne. Eine Einführung in die physikalischen Konzepte* (Springer, Berlin, 2004)
40. V. Holý, T. Baumbach, Nonspecular X-ray reflection from rough multilayers. Phys. Rev. B **49**, 10668 (1994)
41. R. Pynn, Neutron scattering by rough surfaces at grazing incidence. Phys. Rev. B **45**, 602 (1991)
42. S.K. Sinha, X-ray diffuse-scattering as a probe for thin-film and interface structure. J. Phys. III **4**, 1543 (1994)
43. S.K. Sinha, E.B. Sirota, S. Garoff, X-ray and neutron scattering from rough surfaces. Phys. Rev. B **38**, 2297 (1988)
44. D.A. Doshi, E.B. Watkins, J.N. Israelachvili, J. Majewski, Reduced water density at hydrophobic surfaces: effect of dissolved gases. Proc Natl Acad Sci USA **102**, 9458 (2005)

45. F. Rehfeldt, R. Steitz, R.v. Klitzing, S.P. Armes, A.P. Gast, M. Tanaka, Reversible activation of Diblock copolymer monolayers at the interface by pH modulation (2): membrane interactions at the solid/liquid interface. J. Phys. Chem. B **110**, 9177 (2006)
46. T. Schubert, P. Seitz, E. Schneck, M. Nakamura, M. Shibakami, S.S. Funari, O. Konovalov, M. Tanaka, Structure of synthetic transmembrane lipid membranes at the solid/liquid interface studied by specular X-ray reflectivity. J. Phys. Chem. B **112**, 10041 (2008)
47. I.M. Tidswell, B. Ocko, P.S. Pershan, S.R. Wasserman, G.M. Whitesides, J.D. Axe, X-ray specular reflection studies of silicon coated by organic monolayers (alkylsiloxanes). Phys. Rev. B **41**, 1111 (1990)
48. L. Névot, P. Croce, Characterization of surfaces by grazing X-ray reflection—application to the study of polishing of some silicate glasses. Rev. Phys. Appl. **15**, 761 (1980)
49. L.G. Parratt, Surface studies of solids by total reflection of X-rays. Phys. Rev. **95**, 359 (1954)
50. F. Abelès, Recherches théoriques sur les propriétés optiques des lames minces. J. Phys. Rad. **11**, 307 (1950)
51. V. Nitz, M. Tolan, J.-P. Schlomka, O.H. Seeck, J. Stettner, W. Press, M. Stelzle, E. Sackmann, Correlations in the interface structure of Langmuir–Blodgett films observed by X-ray scattering. Phys. Rev. B **54**, 5038 (1996)
52. J.-P. Schlomka, M. Tolan, L. Schwalowsky, J. Stettner, W. Press, X-ray diffraction from Si/Ge layers: diffuse scattering in the region of total external reflection. Phys. Rev. B **51**, 2311 (1995)
53. B. Demé, M. Dubois, T. Zemb, B. Cabane, Effect of carbohydrates on the swelling of a lyotropic lamellar phase. J. Phys. Chem. **100**, 3828 (1996)
54. N. Lei, C.R. Safinya, R.F. Bruinsma, Discrete harmonic model for stacked membranes—theory and experiment. J. Phys. II **5**, 1155 (1995)

Chapter 3
Materials and Methods

In this chapter, the technical details of the materials, preparation methods, and scattering techniques utilized for the study of membrane-bound saccharides are presented.

3.1 Materials

3.1.1 DPPC and Synthetic Glycolipids

3.1.1.1 DPPC

The structure of chain-deuterated DPPC (1,2-Dipalmitoyl-D62-*sn*-Glycero-3-Phosphatidylcholine, DPPC-D) is shown in Fig. 3.1. The molecule consists of two fully deuterated all-saturated hexadecyl chains connected via ester bonds and a glycerol junction to a zwitterionic phosphatidylcholine head group. DPPC-D was purchased from Avanti Polar Lipids (Alabaster, USA).

3.1.1.2 Gentiobiose lipid and Lac1 lipid

The structures of the synthetic glycolipids Gent and Lac1 are shown in Fig. 3.2. Both molecules consist of two all-saturated hexadecyl chains and a neutral disaccharide head group, connected to the hydrocarbon chains via a glycerol junction. Gent possesses a gentiobiose head group (*O*-(*β-D-glucopyranosyl*)-(1→6)-*β-D-glucopyranoside*), which is bent with respect to the molecular axis. In contrast, Lac1 has a cylindrical mono-lactose (*O*-(*β-D-galactopyranosyl*)-(1→4)-*β-D-glucopyranoside*) head group. To optimize the scattering length density

E. Schneck, *Generic and Specific Roles of Saccharides at Cell and Bacteria Surfaces,* Springer Theses, DOI: 10.1007/978-3-642-15450-8_3,
© Springer-Verlag Berlin Heidelberg 2011

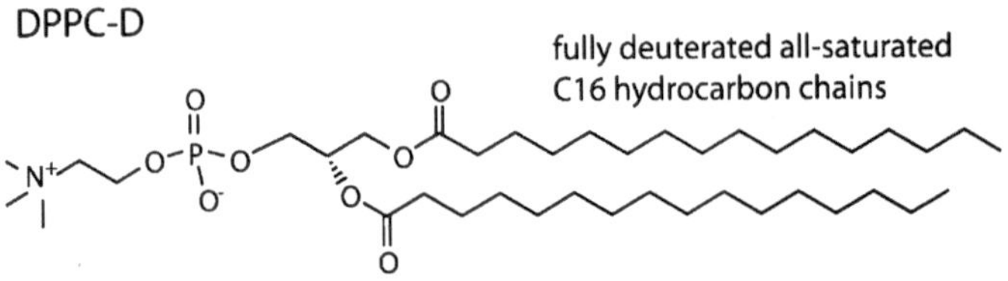

DPPC-D

fully deuterated all-saturated
C16 hydrocarbon chains

Fig. 3.1 Chemical structure of the chain-deuterated deuterated DPPC-D molecule

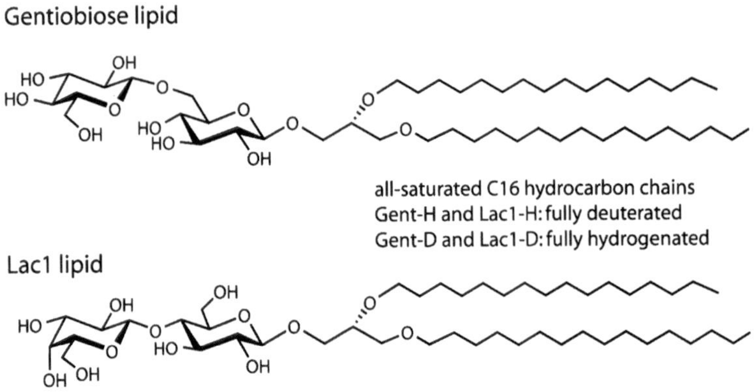

Gentiobiose lipid

all-saturated C16 hydrocarbon chains
Gent-H and Lac1-H: fully deuterated
Gent-D and Lac1-D: fully hydrogenated

Lac1 lipid

Fig. 3.2 Chemical structures of the synthetic Gentiobiose lipid and Lac1 lipid

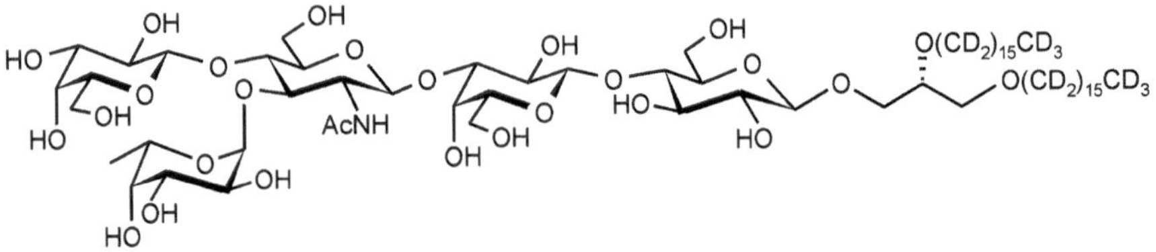

$O(CD_2)_{15}CD_3$
$O(CD_2)_{15}CD_3$

Fig. 3.3 Chemical structure of the synthetic Le^X lipid

contrast in neutron scattering experiments, molecules with fully hydrogenated
hydrocarbon chains (Gent-H, Lac1-H) and with fully deuterated hydrocarbon
chains (Gent-D, Lac1-D) were used. The glycolipids were synthesized by C. Gege
(University of Konstanz, Konstanz, Germany). Details of the syntheses are given
elsewhere [1–3].

3.1.1.3 Le^X lipid

The structure of the synthetic Le^X lipid is shown in Fig. 3.3. The molecule consists
of two fully deuterated all-saturated hexadecyl chains and a neutral pentasaccharide head group which comprises the LewisX trisaccharide D-Gal$\beta(1{\rightarrow}4)$

[L-Fucα(1→3)]D-GlcNAcβ(1→R) connected to the hydrocarbon chains via a mono-lactose unit and a glycerol junction. Le^X lipid was synthesized by C. Gege (University of Konstanz, Konstanz, Germany). Details of the synthesis are given elsewhere [1, 4, 5].

3.1.2 Lipopolysaccharides

The structure of the here studied Lipopolysaccharides is presented in Fig. 3.4. Lipid A is the fundamental building block of all LPS molecules. It consists of two negatively charged phosphorylated glucosamines, bound to six hydrocarbon chains. A certain fraction contains further substitutions, such as an extra palmitoyl chain, a 4-amino-deoxyarabinose, and a phosphatidylethanolamine [6] (Fig. 3.4, indicated in dark grey). The rough mutant LPS Re possesses two more negatively charged 2-keto-3-deoxyoctonoic acid (KDO) units, constituting the "inner core". In addition to that, the rough mutant LPS Ra possesses eight more saccharide units, two of which are phosphorylated [7]. This moiety is known as "outer core". The previously reported values [8] of molecular weight, maximum number of negative charges, and chain melting temperature T_m of Lipid A and the rough mutants LPS Re and LPS Ra are summarized in Table 3.1.

In addition to the largely invariant rough mutant structure, a variable fraction of LPS molecules possesses O-polysaccharides (O-sidechains) in form of repetitive oligosaccharide motives [9]. *P. aeruginosa* produces two different types of O-sidechains: The A-band polysaccharide is a polymer of uncharged D-rhamnose trisaccharides [10], while the B-band is generally built from charged di- to pentasaccharide units. The strain dps 89 (A-B+ mutant of PAO1 LPS) [11, 12] lacks the A-band, and its negatively charged B-band polysaccharide is a polymer of trisaccharide repeat units: (2-acetamido-2,6-dideoxy-D-galactose (N-acetyl-D-fucosamine), 2,3-diacetamido-2,3-dideoxy-D-mannuronic acid, and 3-acetamidi-no-2acetamido-2,3-dideoxy-D-mannuronic acid) [13]. Each repeat unit carries two

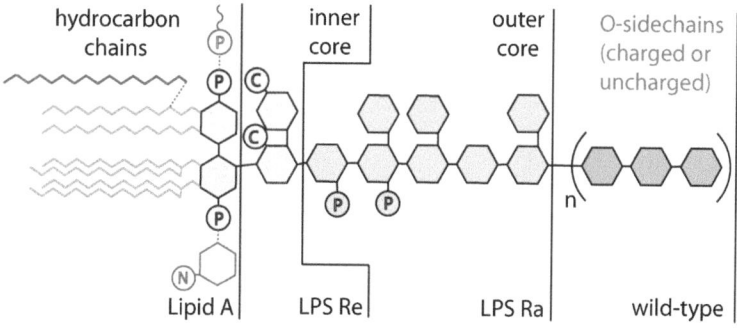

Fig. 3.4 Schematic structures of the studied LPS molecules. Saccharide units are indicated by hexagons and N, P, and C denote amino, phosphate and carboxylate groups, respectively

Table 3.1 Physical characteristics (molecular weight, maximum number of negative charges, and chain melting temperature) of Lipid A the rough mutants LPS Re and LPS Ra

Molecule	Molecular weight (g / mol)	Maximum number of negative charges	Bilayer chain melting temperature (°C)
Lipid A	1797	2	46
LPS Re	2237	4	30
LPS Ra	3835	6	36

negative charges at environmental pH. The length of the B-band polysaccharide can reach beyond 50 repeat units [14, 15] but ranges typically from 20 to 50 repeat units [16]. However, the main fraction of molecules carries either no O-sidechain or only one repeat unit (called "semi-rough" LPS). *Pseudomonas aeruginosa* is involved in respiratory tract infections and is often associated with chronic infections in cystic fibrosis patients [17].

Lipid A and rough mutant LPS (LPS Re and LPS Ra) were kindly provided by U. Seydel and K. Brandenburg (Forschungszentrum Borstel, Borstel, Germany): LPS Re was purified from the bacterial strain R595 of *Salmonella enterica* sv. Minnesota, or from the strain F515 of E. coli. LPS Ra was purified from the bacterial strain *R60* of *Salmonella enterica* sv. Minnesota. Lipid A is a hydrolysis product of LPS Re F515. Details of the purification are given elsewhere [18].

LPS from *P.aeruginosa* dps 89 was isolated and Characterized by E. Papp-Szabo. The isolation was performed using the hot phenol-water method [19] with several modifications [20]. The samples were characterized with SDS-PAGE, Western immunoblotting (developed with monoclonal antibodies specific to A-band and B-band), 1H NMR spectroscopy, and sugar composition analysis. The protein contamination of the material was checked by BCA protein assay (PIERCE, Rockford, USA) and Lowry total protein assay (Peterson's modification, Sigma-Aldrich, Saint Louis, USA) and found to be below 1% w/w. There was no detectable DNA and RNA contamination in the LPS sample. The characterization results [20] suggested that, apart from the fact that A-band O-sidechains are missing, the composition the studied A-B+ LPS is comparable to that of PAO LPS [21]: About 10% of the LPS molecules possess an B-band O-sidechain with 20–50 repeating units, about 70% are rough, and about 20% are semi-rough LPSs.

3.1.3 Chemicals and Buffers

H_2O was double de-ionized with a specific resistance greater than 18 $M\Omega cm$ (MilliQ, Molsheim, France). D_2O was purchased from Euriso-Top (Saint-Aubin, France). All other chemicals were purchased from Fluka (Taufkirchen, Germany) and used without further purification.

Throughout this study several buffers were used, prepared either from 100% H_2O or from 100% D_2O. All buffers contained 5 mM Hepes and were titrated to a pH value of 7.4.

3.1.3.1 Calcium-Free Buffers

"*Ca-free NaCl buffer*" additionally contained 100 mM NaCl.
"*Ca-free KCl buffer*" additionally contained 100 mM KCl.

3.1.3.2 Calcium-Loaded Buffers

"*Ca01 NaCl buffer*" additionally contained 100 mM NaCl and 1 mM $CaCl_2$.
"*Ca02 NaCl buffer*" additionally contained 100 mM NaCl and 2 mM $CaCl_2$.
"*Ca05 NaCl buffer*" additionally contained 100 mM NaCl and 5 mM $CaCl_2$.
"*Ca50 NaCl buffer*" additionally contained 100 mM NaCl and 50 mM $CaCl_2$.
"*Ca50 KCl buffer*" additionally contained 100 mM KCl and 50 mM $CaCl_2$.

For neutron scattering experiments both H_2O-based and D_2O-based buffers were used, depending on the sample, to achieve the maximum contrast in scattering length density. For all other experiments the H_2O buffers were used.

3.2 Preparation Methods

3.2.1 Preparation of Solutions/Suspensions

3.2.1.1 DPPC and Synthetic Glycolipids

DPPC-D and synthetic glycolipids (Gent-H, Gent-D, Lac1-H, Lac1-D, and Le^X lipid) were dissolved in 7:3 mixtures (v/v) of chloroform and methanol at concentrations of 1 or 2 mg/mL. DPPC-D/Le^X lipid mixtures were prepared by mixing the pure solutions corresponding to defined molar fractions of Le^X lipid (2 mol%, 10 mol%, 25 mol% Le^X lipid).

3.2.1.2 Lipid A and Rough Mutant LPS

Lipid A and LPS Re were dissolved in 7:3 mixtures (v/v) of chloroform and methanol at a concentration of 2 mg/mL. Since LPS Ra cannot easily be dissolved in organic solvents, it was suspended in pure D_2O at a concentration of 2 mg/mL and sonified with a tip sonifier (Misonix, Newtown, USA) under mild conditions (interval mode: 0.5 s/0.5 s on/off, constant cooling to $T < 50\ °C$).

3.2.1.3 PAOLPS

PAO1 LPS dps 89 was suspended in *Ca-free KCl buffer* at a concentration of 5 mg/mL. After sonication with a tip sonifier (Misonix, Newtown, USA) under mild conditions (interval mode: 0.5 s/0.5 s on/off, constant cooling to $T < 50\ °C$), the LPS suspension was diluted with *Ca-free KCl buffer* to a final concentration of 1 mg/mL.

3.2.2 Preparation of Solid-Supported Membrane Multilayers

A 0.5 or 1 mL portion of solution / suspension was deposited onto a rectangular (55 mm × 25 mm) Si(100)-substrate with native oxide (Si-Mat, Landsberg am Lech, Germany), which was cleaned by a modified RCA method [22]. During the process of solvent evaporation the amphiphilic molecules self-assemble into planar membrane stacks, aligned parallel with the substrate surface (Fig. 3.5).

To remove residual solvent, the wafers were stored at 70 °C for 3 h, and subsequently in a vacuum chamber overnight. The average number of membranes

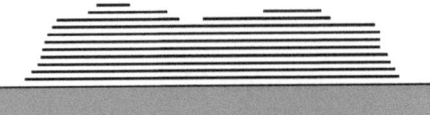

Fig. 3.5 Sketch of a silicon chip coated with membrane multilayers

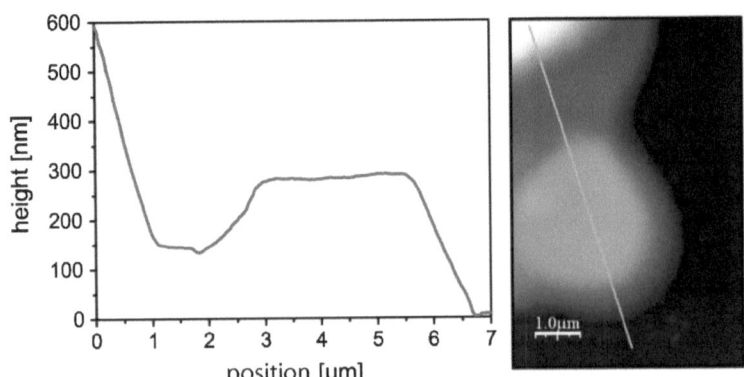

Fig. 3.6 Height profile (*left*) of solid-supported Gent-D membrane multilayers under ambient conditions and original atomic force microscopy (AFM) image (*right*). The *straight line* corresponds to the profile

in the stacks was at the order of several hundred, as can be calculated from the amount of solution and the coated area. Figure 3.6 shows the height profile of solid-supported Gent-D multilayers measured by atomic force microscopy (AFM) under ambient conditions. The profile suggests an estimated in-plane length scale of the solid-supported multilayer patches at the order of several micrometers. To cancel the thermal history of the samples, at least two heating/cooling cycles between 20 and 70 °C were repeated at a high relative humidity ($h_{rel} > 95$ %) and the samples were stored at 4 °C overnight prior to the measurements.

3.2.3 Preparation of Rough Mutant LPS Monolayers at the Air/ Water Interface

To create monolayers at the air/water interface, organic LPS rough mutant solution was droplet-wise deposited onto the buffer surface in a Langmuir trough. *Ca-free NaCl buffer* or *Ca50 NaCl buffer* was used for GIXOS measurements, while *Ca-free KCl buffer* or *Ca50 KCl buffer* was used for X-ray fluorescence measurements. Prior to compression, 20 min time were allowed for the complete evaporation of the solvent. For X-ray measurements the film was compressed to a lateral pressure of 20 mN/m using the home-made Langmuir film balance provided by the beamline ID10B (ESRF, Grenoble, France). Pressure-area isotherms were recorded by monitoring the lateral pressure while compressing the monolayers at a constant rate of 10 Å [2] per molecule per minute using a commercial Langmuir film balance (Kibron, Espoo, Finland).

3.2.4 Preparation of Solid-Supported PAOLPS Monolayers

A rectangular (25 mm × 20 mm) Si(100)-substrate with native oxide (Si-Mat, Landsberg/Lech, Germany), was cleaned by a modified RCA method [22]. A self-assembled monolayer of octadecyltrimethoxysilane (ODTMS, purchased from ABCR, Karlsruhe, Germany) was covalently grafted onto the substrate according to Hillebrandt and Tanaka [23]. Subsequently, the substrate was stored at 70 °C for 3 h and in a vacuum chamber overnight to remove residual solvent molecules. Finally, the functionalized substrate was inserted into a self-built liquid cell for under-water X-ray measurements (see Sect. 3.3.3.3). The LPS suspension was injected into the liquid cell and incubated at 50 °C for 1 h. After extensive rinsing, the sample was subjected to the reflectivity measurement in *Ca-free KCl buffer*. In the next step, the liquid cell was rinsed with an excess of *Ca50 KCl buffer* for the reflectivity measurements in the presence of Ca^{2+} ions.

3.3 Scattering Techniques

3.3.1 X-Ray Scattering

3.3.1.1 Specular X-Ray Reflectivity Experiments

Specular X-ray reflectivity experiments were carried out at the ID10B beamline at ESRF, (Grenoble, France). The scattering geometry is illustrated in Fig. 3.7. A monochromatic beam with a photon energy of 22 keV ($\lambda = 0.56$ Å) was collimated to a vertical beam aperture of 20 μm. At each angle of incidence α_i, corresponding to the momentum transfer perpendicular to the interface $q_z = (4\pi/\lambda)$ sin α_i, the reflectivity was corrected for the beam footprint and for the beam intensity (via an in-beam monitor).

3.3.1.2 GIXOS Experiments at the Air/Water Interface

GIXOS experiments were carried out at the ID10B beamline at ESRF (Grenoble, France). The scattering geometry is illustrated in Fig. 3.8. A monochromatic beam

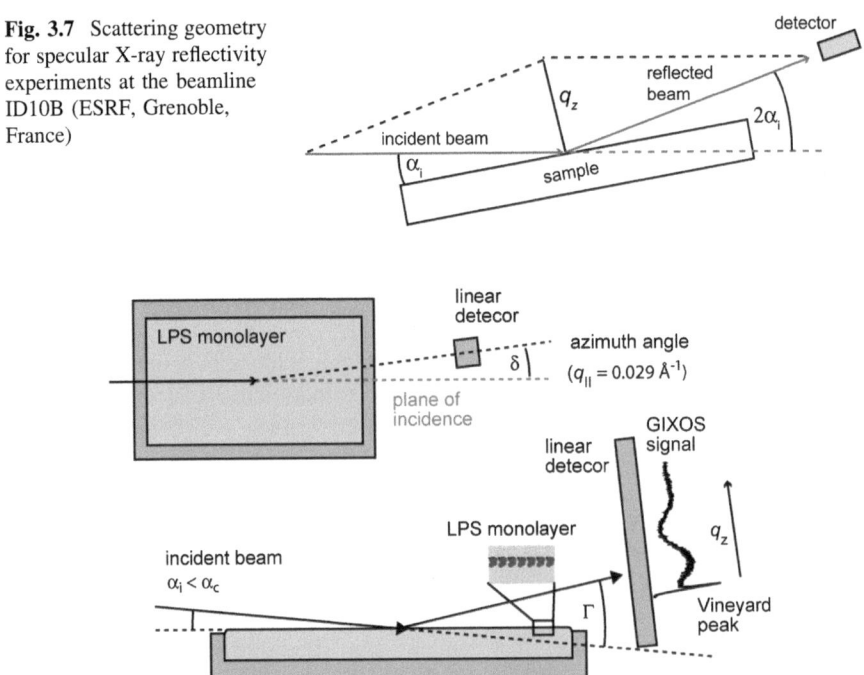

Fig. 3.7 Scattering geometry for specular X-ray reflectivity experiments at the beamline ID10B (ESRF, Grenoble, France)

Fig. 3.8 Scattering Geometry for GIXOS experiments at the beamline ID10B (ESRF, Grenoble, France)

(8 keV, $\lambda = 1.55$ Å) illuminates the monolayer at an incident angle ($\alpha_i = 0.119°$) slightly below the critical angle of the air/water interface α_c.

In the GIXOS (Grazing Incidence X-ray scattering Out of the Specular plane) measurements, the intensity of the scattered beam is collected with a position sensitive linear detector perpendicular to the monolayer surface at an azimuth angle δ near the incidence plane ($q_\parallel = 0.029$ Å^{-1}). In case the in-plane momentum transfer $q_\parallel$ is very small and the topological interface roughnesses are conformal (i.e., fully correlated, see Sect. 2.2.3), the measured diffuse intensity is connected to the corresponding reflectivity curve [24, 25]:

$$I(q_z) \propto |T(k_{out})|^2 \frac{R(q_z)}{R_F(q_z)},$$

$$q_z = \frac{2\pi}{\lambda}[\sin(\Gamma - \alpha_i) + \sin\alpha_i].$$

Here, $I(q_z)$ denotes the intensity measured in a GIXOS experiment, $R(q_z)$ is the corresponding specular reflectivity (see Sect. 2.2.2) as measured in a "$\theta - 2\theta$" scan, $R_F(q_z)$ denotes the Fresnel reflectivity from an ideal surface and $T(k_{out})$ represents the transmission function for the grazing incidence configuration. GIXOS data were fitted using slab models for the vertical electron density profile $\rho(z)$. The calculations were based on the Master-Formula for specular reflectivity (see Sect. 2.2.2):

$$R(q_z) \propto R_F(q_z)\left|\int \frac{d\rho(z)}{dz}\exp(iq_z z)dz\right|^2$$

During X-ray irradiation the monolayer was kept in helium atmosphere to avoid undesirable scattering in air.

3.3.1.3 X-Ray Fluorescence Experiments

X-ray fluorescence experiments were carried out at the ID10B beamline at ESRF, (Grenoble, France). The scattering geometry is illustrated in Fig. 3.9. A monochromatic X-ray beam (8 keV, $\lambda = 1.55$ Å) hits the air/water interface at incident angles α_i below and above the critical angle of total reflection of the air/water

Fig. 3.9 Scattering Geometry for X-ray fluorescence experiments at the beamline ID10B (ESRF, Grenoble, France). Fluorescence radiation is indicated with *arrows*

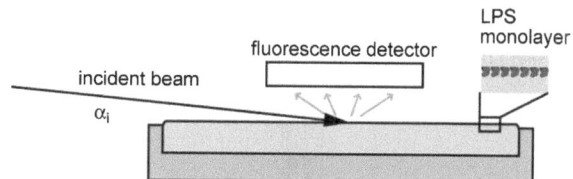

Fig. 3.10 (left) X-ray fluorescence spectra from an LPS Re monolayer on *Ca50 KCl buffer* recorded at two different angles of incidence α_i, below and above the critical angle of total reflection

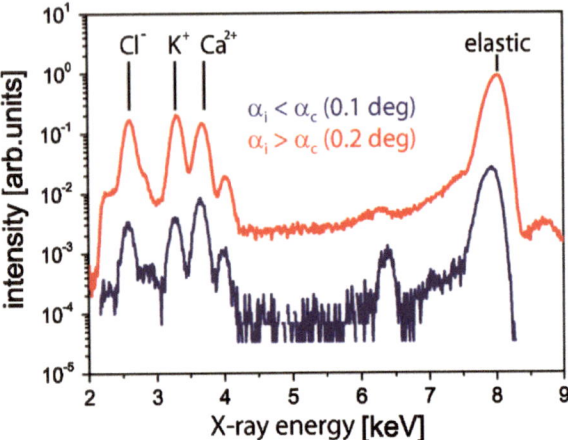

interface α_c. At each angle, the fluorescence radiation emitted by the ions was recorded with an energy sensitive detector.

The signals were normalized with respect to detector counting efficiency. The resulting spectra (see Fig. 3.10 left) were fitted with multiple Gaussian profiles to extract the total fluorescence intensity of each element. α_i values were transformed into q_z values (i.e., the component of the scattering vector normal to the interface $q_z = (4\pi/\lambda)\sin \alpha_i$) for direct comparison with modeled fluorescence signals. The recorded fluorescence intensities were normalized on the elastically scattered intensity (i.e. divided by the 8 keV peak intensity) to compensate for any systematic differences between the experiments, such as absorption by the sample environment and illumination properties. In the last step the fluorescence signals from the monolayer systems were normalized through dividing them by the corresponding signals recorded from blank buffer. This procedure avoids all difficulties that may arise when dealing with the exact description of the experimental geometry, such as footprint size and fluorescence aperture considerations, as well as with the absorption of the illumination and the fluorescence emission in the bulk media, as these effects apply equally to all experiments (see Sect. 4.3). During X-ray irradiation the monolayer was kept in helium atmosphere to avoid undesirable scattering from air molecules and to minimize the undesired X-ray fluorescence of argon.

3.3.2 Neutron Scattering

Neutron scattering experiments were carried out at the D16 membrane diffractometer of the Institut Laue-Langevin (ILL, Grenoble, France). Figure 3.11 shows the geometry of the experiment (top view). A monochromatic neutron beam ($\Delta\lambda/\lambda = 1\%$) of $\lambda = 4.54$ Å or $\lambda = 4.73$ Å reaches the sample through the

Fig. 3.11 Scattering geometry for specular and off-specular neutron scattering experiments at the instrument D16 (ILL, Grenoble, France)

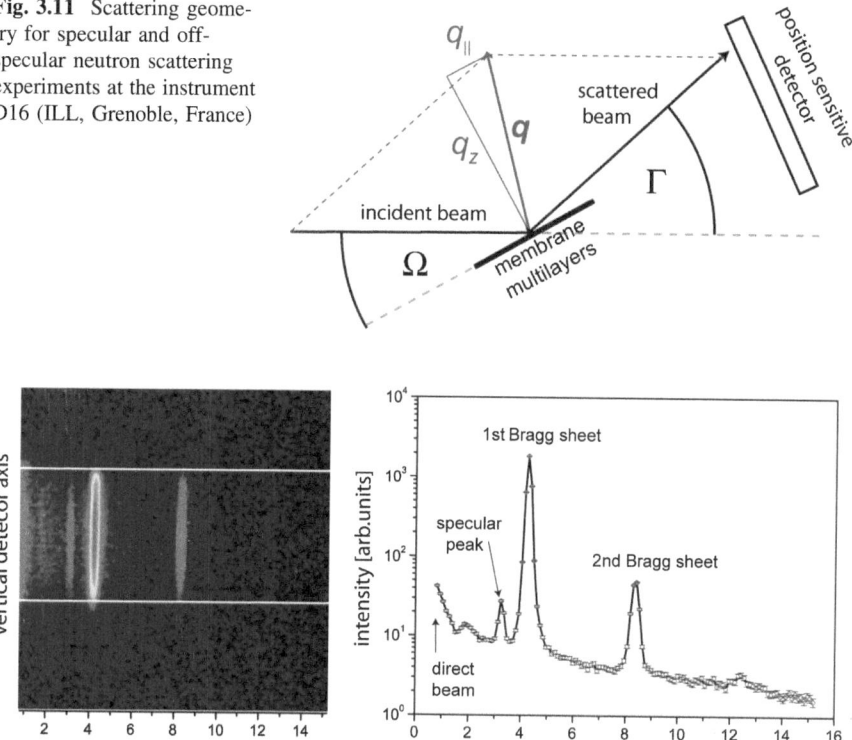

Fig. 3.12 (*left*) D16 detector readout for $\Omega = 1.6°$. *Horizontal white lines* indicate the integration range. (*right*) Integrated intensity plotted as a function of Γ showing the specular reflection and two Bragg sheets

aluminum windows of the sample chamber, while the incident angle Ω (i.e., the angle between the incident beam and the sample plane) is adjusted by a rotation stage.

The intensity of the beam diffracted from the sample is recorded by a position sensitive ^{3}He 2D detector with 128×128 channels. Γ denotes the angle between the scattered and the incident beam. The sample was rotated stepwise with respect to the incident beam. The beam width was 2 mm horizontally and 25 mm vertically. For each measurement at an angle Ω, the detector readout was normalized to the intensity of the incident beam (via an in-beam monitor), the channel sensitivity, and the illuminated sample area.

Subsequently, each 2D detector readout was integrated in the vertical direction, resulting in a one-dimensional intensity projection as a function of the horizontal detector channel position (corresponding to a Γ value). This is shown in Fig. 3.12 for $\Omega = 1.6°$. Thus, one Ω-scan yielded the recorded intensity as a function of Ω and Γ (see Fig. 3.13 left). The datasets in angular coordinates can be transformed into reciprocal space maps (Fig. 3.13 right) by geometrical considerations:

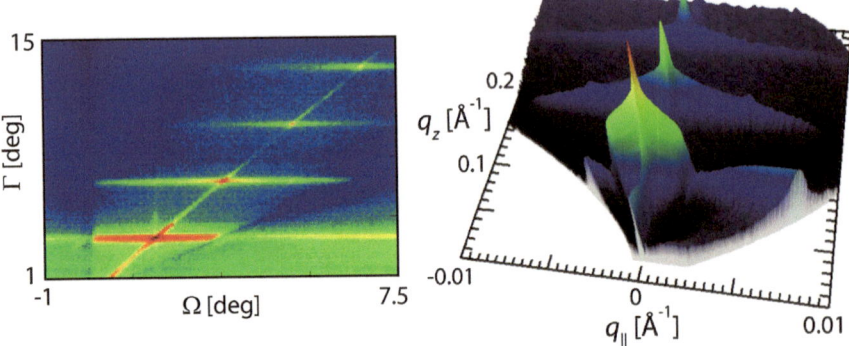

Fig. 3.13 (*left*) Angular intensity map and (*right*) corresponding reciprocal space map (RSM)

$$q_z = \frac{2\pi}{\lambda}[\sin(\Gamma - \Omega) + \sin(\Omega)],$$

$$q_{\parallel} = \frac{2\pi}{\lambda}[\cos(\Gamma - \Omega) - \cos(\Omega)].$$

Here, q_z and $q_{\parallel}$ denote the scattering vector components perpendicular and parallel to the sample plane (see Figs. 3.11 and 2.12).

3.3.3 Sample Environments

3.3.3.1 Humidity Chamber for Neutron Scattering under Vapor Conditions

For measurements at controlled temperature and relative humidity, the humidity chamber provided by the ILL was used. A sketch is presented in Fig. 3.14. The humidity chamber consists of two cells: a sample chamber and a chamber containing a reservoir for liquid water (H_2O or D_2O). The cells are thermally isolated but interconnected for vapor exchange. Two independent thermostats (Phoenix II, Haake, Karlsruhe, Germany, $\Delta T = 0.1\ °C$), controlled by the instrument control program, regulate the temperatures in the sample chamber T_s and the water reservoir T_r, which allows for the regulation of both sample temperature and relative humidity (i.e., the osmotic pressure exerted to the sample, defined by the chemical potential of water vapor in equilibrium) in the chamber. The relative humidity in the sample chamber, h_{rel}, is determined by T_r and T_s:

$$h_{rel}(T_r) = p(T_r)/p(T_s)$$

$p(T)$ denotes the saturation water vapor pressures as can be calculated from a Taylor series with the known parameters for H_2O and D_2O. To ensure equilibration, the samples were kept at each temperature and humidity condition for at least 30 min prior to the measurement.

Fig. 3.14 Sketch of the ILL
setup for temperature and
humidity control used for
neutron scattering
experiments

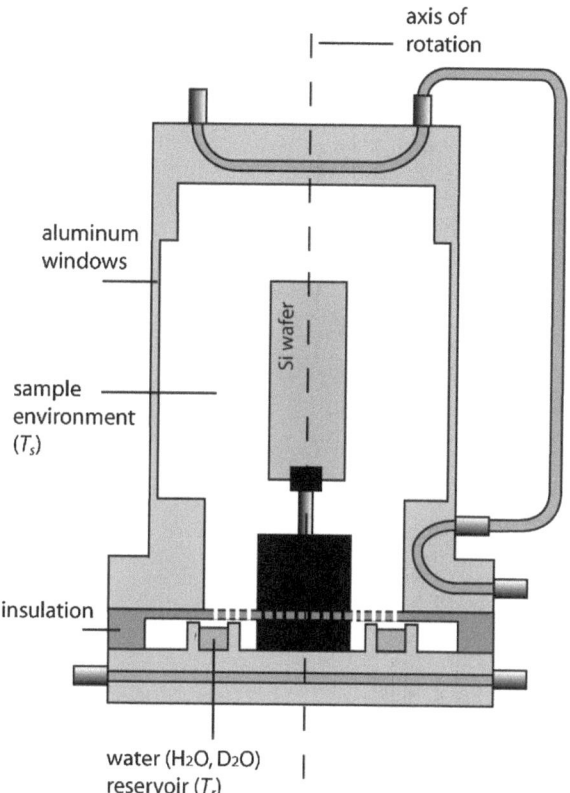

To achieve the maximum contrast in scattering length density between hydrated
head groups and hydrocarbon chains, chain-deuterated samples were hydrated by
H_2O vapor, while hydrogenated samples were hydrated by D_2O vapor.

3.3.3.2 Liquid Cell for Neutron Scattering under Bulk Water Conditions

For neutron scattering experiments under bulk water conditions a self-built liquid
cell as shown in Fig. 3.15 was used. The liquid cell consists of two rectangular Si-
wafers (55 mm × 25 mm), one of which is coated with the membrane multilayers.
The wafers are separated by glass slide pieces (thickness: 0.10 mm), and the
capillary force confines a thin layer of water or aqueous buffer between the two
wafers. During measurements, the entire liquid cell is placed in the climate
chamber at controlled temperature and high relative humidity (>95%) to minimize
the evaporation of water. To achieve the maximum contrast in scattering length
density between hydrated head groups and hydrocarbon chains, chain-deuterated
samples were hydrated by H_2O or H_2O-based buffers, while hydrogenated samples
were hydrated by D_2O or buffers containing 100% D_2O.

Fig. 3.15 Sketch of the self-developed liquid cell. The solid supported membrane multilayers are immersed in a thin layer of bulk aqueous medium between the two silicon wafers. The neutron beam reaches the sample through the silicon support

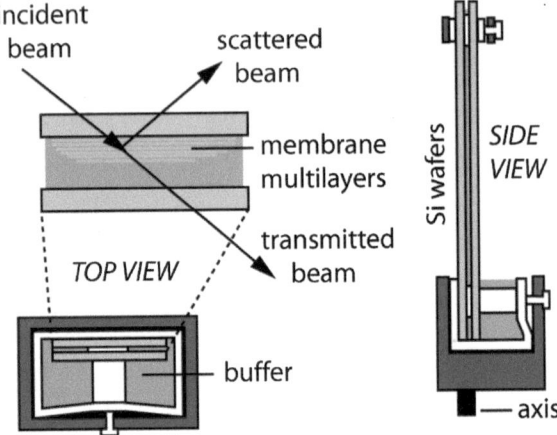

Fig. 3.16 Sketch of the self-built liquid cell for high-energy X-ray reflectivity measurements

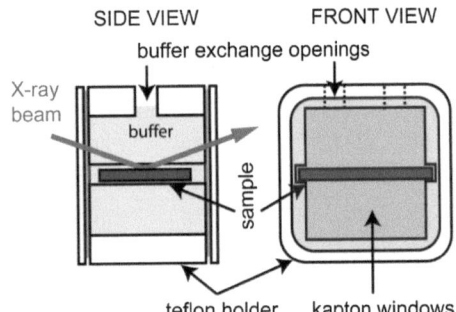

3.3.3.3 Liquid Cell for High-Energy X-Ray Reflectivity Measurements

For high-energy X-ray reflectivity measurements a self-built liquid cell was used. Its design was inspired by the work of Miller et al. [26, 27]. Figure 3.16 shows a sketch of the cell. It consists of a PTFE main body, into which a silicon chip (24 mm × 20 mm) can be placed. The X-ray beam reaches the sample through Kapton windows. Two holes in the top side allow for the in situ buffer exchange.

References

1. C. Gege, J. Vogel, G. Bendas, U. Rothe, R.R. Schmidt, Synthesis of the Sialyl Lewis X Epitope attached to glycolipids with different core structures and their selectin-binding characteristics in a dynamic test system. Chem. Eur. J. **6**, 111 (2000)
2. M.F. Schneider, G. Mathe, M. Tanaka, C. Gege, R.R. Schmidt, Thermodynamic properties and swelling behavior of glycolipid monolayers at interfaces. J. Phys. Chem. B **105**, 5178 (2001)

3. M. Tanaka, S. Schiefer, C. Gege, R.R. Schmidt, G.G. Fuller, Influence of subphase conditions on interfacial viscoelastic properties of synthetic lipids with gentiobiose head groups. J. Phys. Chem. B **108**, 3211 (2004)

4. C. Gege, A. Geyer, R.R. Schmidt, Synthesis and molecular tumbling properties of sialyl Lewis X and derived neoglycolipids. Chem. Eur. J. **8**, 2454 (2002)

5. K. Saxena, R.I. Duclos, P. Zimmermann, R.R. Schmidt, G.G. Shipley, Structure and properties of totally synthetic galacto- and gluco-cerebrosides. J. Lipid Res. **40**, 839 (1999)

6. O. Lüderitz, M. Freudenberg, C. Galanos, V. Lehmann, E.T. Rietschel, D.H. Shaw, Lipopolysaccharides of gram-negative bacteria. Curr. Top. Membr. Transp. **17**, 79 (1982)

7. K. Brandenburg, J. Andrä, M. Müller, M.H.J. Koch, P. Garidel, Physicochemical properties of bacterial glycopolymers in relation to bioactivity. Carbohydr. Res. **338**, 2477 (2003)

8. K. Brandenburg, U. Seydel, Physical aspects of structure and function of membranes made from lipopolysaccharides and free lipid A. Biochim. Biophys. Acta **775**, 225 (1984)

9. M. Caroff, D. Karibian, Structure of bacterial lipopolysaccharides. Carbohydr. Res. **338**, 2431 (2003)

10. T.L. Arsenault, D.W. Hughes, D.B. MacLean, W.A. Szarek, A.M.B. Kropinski, J.S. Lam, Structural studies on the polysaccharide portion of 'A-band' lipopolysaccharide from a mutant (AK1401) of Pseudomonas aeruginosa strain PAO1. Can. J. Chem. **69**, 1273 (1991)

11. J.L. Kadurugamuwa, J.S. Lam, T.J. Beveridge, Interaction of gentamicin with the A band and B band lipopolysaccharides of Pseudomonas aeruginosa and its possible lethal effect. Antimicrob. Agents Chemother. **37**, 715 (1993)

12. J. Lightfoot, J.S. Lam, Molecular cloning of genes involved with expression of A-band lipopolysaccharide, an antigenically conserved form, in Pseudomonas aeruginosa. J. Bacteriol. **173**, 5624 (1991)

13. Y.A. Knirel, E.V. Vinogradov, N.A. Kocharova, N.A. Paramonov, N.K. Kochetkov, B.A. Dmitriev, E.S. Stanislavsky, B. La'nyi, O-specific polysaccharides and serological classification of Pseudomonas aeruginosa. Acta Microbiol. Hung. **35**, 3 (1988)

14. J.S. Lam, L.L. Graham, J. Lightfoot, T. Dasgupta, T.J. Beveridge, Ultrastructural examination of the lipopolysaccharides of Pseudomonas aeruginosa strains and their isogenic rough mutants by freeze-substitution. J. Bacteriol. **174**, 7159 (1992)

15. I. Sadovskaya, J.R. Brisson, P. Thibault, J.C. Richards, J.S. Lam, E. Altman, Structural characterization of the outer core and the O-chain linkage region of lipopolysaccharide from Pseudomonas aeruginosa serotype O5. Eur. J. Biochem. **267**, 1640 (2000)

16. C. Daniels, C. Griffiths, B. Cowles, J.S. Lam, Pseudomonas aeruginosa O-antigen chain length is determined before ligation to lipid A core. Environ. Microbiol. **4**, 883 (2002)

17. N. Høiby, Prevention and treatment of infections in cystic fibrosis. Int. J. Antimicrob. Agents **1**, 229 (1992)

18. C. Galanos, O. Lüderitz, O. Westphal, A new method for the extraction of R lipopolysaccharides. Eur. J. Biochem. **9**, 254 (1969)

19. O. Westphal, K. Jann, Bacterial lipopolysaccharides. Methods Carbohydr. Chem. **5**, 83 (1965)

20. E. Schneck, E. Papp-Szabo, B.E. Quinn, O.V. Konovalov, T.J. Beveridge, D.A. Pink, M. Tanaka, Calcium ions induce collapse of charged o-side chains of lipopolysaccharides from Pseudomonas aeruginosa. J. R. Soc. Interface **6**, S671 (2009)

21. E. Papp-Szabo, unpublished results

22. W. Kern, D.A. Puotinen, Cleaning solutions based on hydrogen peroxide for use in silicon semiconductor technology. RCA Rev. **31**, 187 (1970)

23. H. Hillebrandt, M. Tanaka, Electrochemical characterization of self-assembled alkoxysiloxane monolayers on indium tin oxide semiconductor electrodes. J. Phys. Chem. B **105**, 4270 (2001)

24. S.M. O'Flaherty, L. Wiegart, O. Konovalov, B. Struth, Observation of zinc phthalocyanine aggregates on a water surface using grazing incidence X-ray scattering. Langmuir **21**, 11161 (2005)

25. L. Wiegart, B. Struth, M. Tolan, P. Terech, Thermodynamic and structural properties of phospholipid langmuir monolayers on hydrosol surfaces. Langmuir **21**, 7349 (2005)
26. C.E. Miller, J. Majewski, T. Gog, T.L. Kuhl, Characterization of biological thin films at the solid-liquid interface by X-Ray reflectivity. Phys. Rev. Lett. **94**, 238104 (2005)
27. E. Nováková, K. Giewekemeyer, T. Salditt, Structure of two-component lipid membranes on solid support: an X-ray reflectivity study. Phys. Rev. E **74**, 051911 (2006)

Chapter 4
Theoretical Modeling

This chapter contains theoretical considerations going beyond the established theory and newly developed for this thesis. The results constitute the basis for the interpretation of the experimental results presented in Chaps. 5 and 6. If not stated otherwise, all numerical calculations were performed using the IDL software package (ITT Visual Information Solutions).

4.1 Determination of Mechanical Properties of Interacting Membranes

In Chaps. 5 and 6, the mechanical properties (i.e., the compression modulus B and the membrane bending rigidity κ) of interacting membranes are determined by analyzing specular and off-specular neutron scattering intensities. The procedure developed in this thesis is presented and discussed in the following.

In the first step, realistic membrane displacement correlation functions based on the mechanical parameters are generated. In the second step, the correlation functions are used to model the experimentally obtained two-dimensional scattering intensity maps including specular and off-specular contributions. The combination of these two steps provides a direct connection between the experimentally observed scattering intensities and the underlying mechanical parameters B and κ of the interacting membranes.

4.1.1 Membrane Displacement Correlation Functions

According to Lei et al. [1] in an infinitely expanded set of oriented multilayers the membrane displacement correlation functions $g_k(r) := \left\langle [u_{n \pm k}(\vec{r_0} - \vec{r}) - u_n(\vec{r_0})]^2 \right\rangle_{\vec{r_0}}$

E. Schneck, *Generic and Specific Roles of Saccharides at Cell and Bacteria Surfaces*, Springer Theses, DOI: 10.1007/978-3-642-15450-8_4,
© Springer-Verlag Berlin Heidelberg 2011

can be expressed with the Caillé parameter η and the de Gennes parameter λ, both determined by the mechanical parameters B and κ (see Sect. 2.1.3):

$$g_k(r) = \frac{d^2}{\pi^2}\eta \int_0^\infty f_k dq_{\parallel}, \quad \text{with: } f_k = \frac{\left[1 - J_o(q_{\parallel}r)\exp\left(-\lambda k q_{\parallel}^2 d\right)\right]}{q_{\parallel}\sqrt{1 + \frac{\lambda^2 d^2}{4}q_{\parallel}^4}}, \quad r = |\vec{r}|$$

Here, $q_{\parallel}$ is inversely proportional to the fluctuation wavelength and denotes the reciprocal space coordinate parallel to the membrane plane. Figure 4.1 (left panel) shows $g_k(r)$ for various values of k. For the calculations, $\eta = 0.04$ and $\lambda = 20$ Å was used. It is seen that $g_k(r)$ diverges with increasing lateral distance r, coinciding with an infinite root mean square (RMS) roughness σ. This is not realistic for membrane multilayers on planar solid substrates, where fluctuations of long wavelengths are suppressed due to the finite sample size. In this thesis, a lower integration limit, $2\pi/R$, is introduced for the calculation of displacement correlation functions, in order to model a system with finite RMS roughness:

$$g_k(r) = \frac{d^2}{\pi^2}\eta \int_{2\pi/R}^\infty f_k dq_{\parallel} \tag{4.1}$$

Lei et al. [1] used an upper integration limit in the scattering intensity calculation to model specular scattering signals from surfactant multilayers, which is mathematically equivalent to the approach shown in this thesis.

Here, the free parameter R, the effective cut-off radius, accounts for two effects:

1. The finite lateral size of the membrane patches acts as an upper limit for the wavelength at which the membranes can fluctuate.
2. Long wavelength membrane fluctuations are more collective (i.e., they involve more membranes, see Sect. 2.1.3) and are therefore damped more strongly by the flat solid support in samples with a finite number of layers [2].

The consequence of the cut-off radius (here: $R = 1$ μm) is illustrated in Fig. 4.1 (right panel), where the curves saturate to the finite value $g_k(r \to \infty) \to 2\sigma^2$. Only this lower integration limit $2\pi/R$ enables to model the full set of realistic membrane displacement correlation functions, necessary for the simulation of two-dimensional reciprocal space maps including specular and off-specular parts. Note that the membrane displacement correlation functions are fully determined by the three free parameters η, λ, and R.

Based on the finite RMS roughness σ, the membrane displacement correlation functions $g_k(r)$ are translated into the corresponding height-height correlation functions $C_k(r)$ (see Sect. 2.1.3):

$$C_k(r) := \langle u_{n\pm k}(\vec{r_0} - \vec{r}) \cdot u_n(\vec{r_0})\rangle_{\vec{r_0}} = \sigma^2 - g_k(r)/2, \quad r = |\vec{r}|$$

The height–height correlation functions are shown in Fig. 4.2.

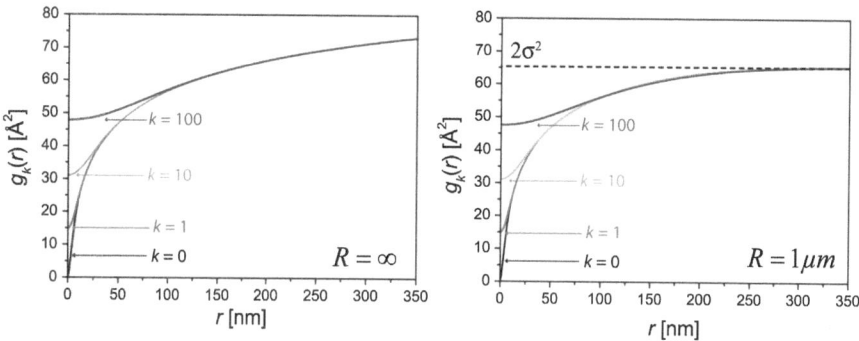

Fig. 4.1 Membrane displacement correlation functions $g_k(r)$ calculated according to Eq. 4.1 for $R = \infty$ (*left*) and $R = 1\ \mu m$ (*right*). The finite cut-off radius $R = 1\ \mu m$ results in a saturation value of the correlations for large r, $g_k(r \to \infty) \to 2\sigma^2$. In the calculations, $\eta = 0.04$ and $\lambda = 20$ Å was used

Fig. 4.2 Height–height correlation functions $C_k(r)$ corresponding to the membrane displacement correlation functions ($\eta = 0.04$, $\lambda = 20$ Å, and $R = 1\ \mu m$) presented in Fig. 4.1 (*left panel*)

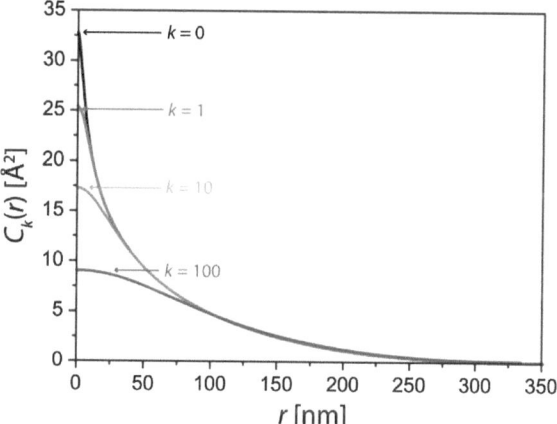

4.1.2 Calculation of Specular and Off-Specular Scattering Signals

In kinematic approximation, the scattering function from a set of stratified interfaces with correlated roughness is given by Sinha [3] (see Sect. 2.2.3):

$$S(q_z, q_{\parallel}) \propto \frac{1}{q_z^2} \sum_{n=1}^{N} \sum_{m=1}^{N} e^{-\frac{1}{2}q_z^2\left(\sigma_n^2 + \sigma_m^2\right)} \Delta\rho_n\, \Delta\rho_m\, e^{-iq_z(z_n - z_m)} \int_{-\infty}^{\infty} e^{q_z^2 C_{nm}(r)}\, e^{-iq_{\parallel}r}\, dr \quad (4.2)$$

N denotes the total number of stratified interfaces, $\Delta\rho_n$ the step in scattering length density across the nth interface, and C_{nm} the cross-correlation function between the nth and the mth interface (see Sect. 2.1.3). In case of membrane multilayers, Eq. 4.2 can be simplified according to their periodicity:

$$\Delta\rho_n = \Delta\rho_m = \Delta\rho, \quad z_n - z_m = (n - m)d$$

It is further plausible to ignore the different behavior of the layers close to the upper and bottom boundaries, since the scattering signals are vastly dominated by the bulk membrane stacks [4, 5]. It can therefore be assumed that all membranes possess the same RMS roughness, and that the correlation functions only depend on the distance between two considered membranes:

$$\sigma_n = \sigma_m = \sigma, \quad C_{nm}(r) = C_{mn}(r) = C_{|n-m|}(r)$$

such that the scattering function simplifies to

$$S(q_z, q_{\parallel}) \propto \frac{e^{-q_z^2\sigma^2}}{q_z^2} \sum_{n=1}^{N}\sum_{m=1}^{N} e^{-i(n-m)q_z d} \int_{-\infty}^{\infty} e^{q_z^2 C_{|n-m|}(r)} e^{-iq_{\parallel}r}dr.$$

As the addends of the double summation only depend on $(n - m)$, the expression can be rewritten into a single summation with the index $j := n - m$,

$$S(q_z, q_{\parallel}) \propto \frac{e^{-q_z^2\sigma^2}}{q_z^2} \left[\sum_{j=-(N-1)}^{N-1} (N - |j|)e^{-ijq_z d} \int_{-\infty}^{\infty} e^{q_z^2 C_{|j|}(r)} e^{-iq_{\parallel}r} dr \right].$$

Here, the identity $\sum_{n=1}^{N}\sum_{m=1}^{N} f(n - m) = \sum_{n-m=-(N-1)}^{N-1} (N - |n - m|)\cdot f(n - m)$ was used.

By introducing $k := |j|$, the expression can be further simplified:

$$S(q_z, q_{\parallel}) \propto \frac{e^{-q_z^2\sigma^2}}{q_z^2} \left[N\int_{-\infty}^{\infty} e^{q_z^2 C_0(r)} e^{-iq_{\parallel}r} dr + 2\sum_{k=1}^{N-1} (N - k)\cos(kq_z d) \int_{-\infty}^{\infty} e^{q_z^2 C_k(r)} e^{-iq_{\parallel}r} dr \right]$$

$$(4.3)$$

Here, $\sum_{j=-(N-1)}^{N-1} (N - |j|)\cdot f(j) = N\cdot f(|j| = 0) + \sum_{|j|=1}^{N-1} (N - |j|)\cdot [f(|j|) + f(-|j|)]$ was used.

The scattering function (see Eq. 4.3) is fully determined by the above derived correlation functions $C_k(r)$, and therefore by the three free parameters η, λ, and R. In contrast to previous approaches, where mechanical properties were calculated either from the power-law decays or from the numerical back-transformations of integrated Bragg sheet intensities [4], the here presented method allows for the complete description of reciprocal space maps by the underlying continuum-mechanical parameters B and κ, and thus for the global comparison with experimentally measured intensities.

The reciprocal space maps (RSMs) described by Eq. 4.3 comprise the specular axis (along $q_{\parallel} \cong 0$), and the Bragg sheets whose shape is characteristic for the correlation functions $C_k(r)$. The Bragg sheets possess intensity maxima where they

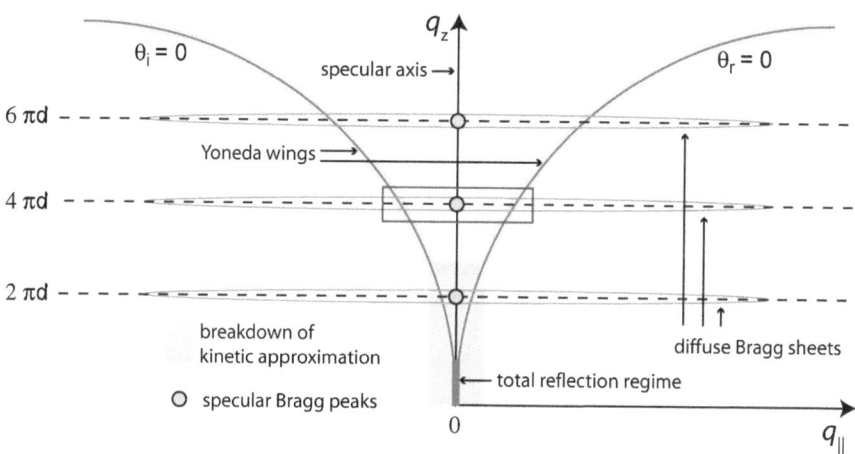

Fig. 4.3 Sketch of a reciprocal space map from solid supported membrane multilayers recorded in the geometry used for neutron scattering experiments (see Sect. 3.3.2 and Fig. 3.11). Features not captured by the used kinematic approximation are highlighted in red. In the vicinity of these features (indicated in *yellow*) the approximation fails

intersect with the specular axis (called "specular Bragg peaks"). These features are schematically illustrated in Fig. 4.3. However, in experiments additional features are observed, which are not captured by the kinematic approximation. They are indicated in red, and comprise the "Yoneda wings" (where refraction effects become relevant) and the region of total reflection. The vicinity of these features and, more generally, regions where strong scattering occurs, has to be omitted while comparing experimental data with simulations in kinematic approximation (see Sect. 2.2). These regions are indicated in yellow, and typically include the first order specular Bragg peak, as shown in the figure. In this work, simulated scattering intensities are compared with experimental data in the vicinity of the second Bragg sheet, roughly as indicated with a blue rectangle. Here, the validity of the kinematic approximation is ensured, except for the close vicinity of the Yoneda wings, which is omitted from the comparison.

To maintain a uniform grid in experimental and simulated data sets, and to be able to perform a proper calculation of measurement error propagation, the simulated and measured scattering intensities are compared in the angular coordinates of the neutron scattering experiments, Γ and Ω (see Sect. 3.3.2, the scattering geometry is shown in Fig. 3.11):

$$S(\Gamma, \Omega) = S(q_z(\Gamma, \Omega), \quad q_{||}(\Gamma, \Omega))$$

For the calculations $N = 100$ was used, sufficient to generate model signals much sharper than the instrumental resolution. The latter was included by convoluting the signal in Γ and Ω directions with Gaussian functions, representing the point spread function of the measurement as it results from the finite angular width and the wavelength spread of the neutron beam.

$$S_C(\Gamma, \Omega) = A(\Gamma, \Omega) \otimes S(\Gamma, \Omega)$$

In practice, the convolution in the Ω direction was achieved using the Fourier convolution theorem in the computation of the Fourier integrals in Eq. 4.3, while the convolution in the Γ direction was applied to the simulated intensity maps. Figure 4.4 presents the comparison of a measured (left) and the corresponding simulated (right) second Bragg sheet in angular coordinates. Ω and Γ ranges roughly correspond to the blue rectangle in Fig. 4.3. The parameters η, λ, and R were chosen such that the best agreement between experimental data and simulation is achieved. It is seen that the global shape of the experimental Bragg sheet is well captured by the simulation. However, for a better visual comparability the following two representations are used in Chaps. 5 and 6 (e.g., Figs. 5.6, 5.13):

1. The Γ-integrated Bragg sheet intensity as a function of $q_{\parallel}$.
2. The fitted Γ-width of the Bragg sheet as a function of $q_{\parallel}$.

While the first representation is characteristic for the membrane displacement self-correlation function $C_0(r)$, the second one is characteristic for the de Gennes parameter λ of a system.[1] Moreover, the two representations are complementary for the description of the Bragg sheets and therefore their combination is well suited to visualize the global agreement between experimental data and simulation.

The accuracy with which the parameters η, λ, and R can be determined from the measured scattering intensities depends on the experimental errors. This is illustrated in Fig. 4.5 for Gent-D in H_2O atmosphere at $T = 80\ °C$ and $h_{rel} \approx 95\ \%$ (see Chap. 5), where the influence of a variation of the parameters from the optimum on the match with the experimental data points is shown. It is seen that for all three parameters, a slight deviation from the optimum results in a significant mismatch with the experimental data points. Furthermore, the nature of the mismatch is seen to be parameter-characteristic, which proves the uniqueness of the parameter set that results in a good match. In the shown case, excellent accuracy is achieved for η and R, while λ is determined to a lesser extent (see Fig. 4.5). However, scattering signals that allow for a precise determination of λ are presented in 5.2.3 (see Fig. 5.13, right column).

---

[1] Salditt et al. [5] have shown that, if the scattering function $S(q_z, q_{\parallel})$ is integrated over one "Brillouin zone" (i.e., from $q_z = (2n-1)\pi/d$ to $q_z = (2n+1)\pi/d$), the contribution of the cross-correlation functions $C_{k \neq 0}(r)$ to the integrated intensity vanishes in good approximation. Furthermore they have shown that the integration range can be narrowed to cover one Bragg sheet. $I_{sheet}^{q_z-int}(q_{\parallel}) := \int_{sheet} S(q_z, q_{\parallel}) dq_z \cong \int_{(2n-1)\pi/d}^{(2n+1)\pi/d} S(q_z, q_{\parallel}) dq_z \propto \int_{-\infty}^{\infty} e^{q_z^2 C_0(r)} e^{-iq_{\parallel}r} dr$. Although several assumptions and approximations have to be taken for this conclusion, the integrated intensity is seen to be surprisingly independent of the cross correlation terms $C_{k \neq 0}(r)$ and can be well represented by solely the self correlation function $C_0(r)$. Due to the tight connection between Γ and q_z, not only $I_{sheet}^{q_z-int}(q_{\parallel})$, but also $I_{sheet}^{\Gamma-int}(q_{\parallel})$ is dominated by $C_0(r)$. On the other hand, the q_z-width of the Bragg sheets, $W_{sheet}^{q_z}(q_{\parallel})$, depends only on the de Gennes parameter λ and is independent of η and R. As motivated above, the same holds for the Γ-width of the Bragg sheets, $W_{sheet}^{\Gamma}(q_{\parallel})$, in good approximation.

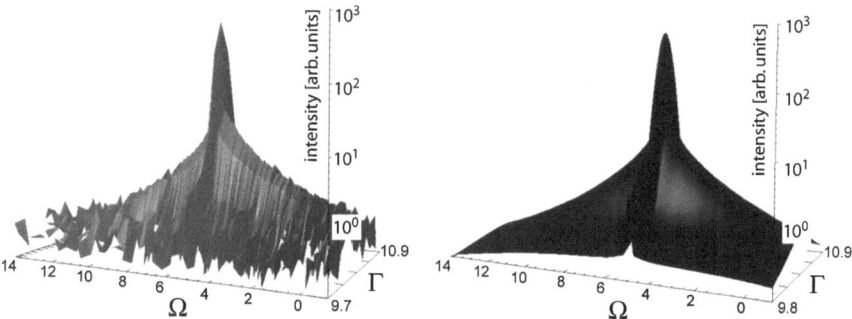

Fig. 4.4 Scattering intensity of the second Bragg sheet as a function of the angles Ω and Γ. System: Gent-D in H_2O atmosphere at $T = 80$ °C and $h_{rel} \approx 95\%$ (see Chap. 5). Ω and Γ ranges roughly correspond to the *blue rectangle* in Fig. 4.3. (*Left*) experimental data. (*Right*) simulation according to Eq. 4.3

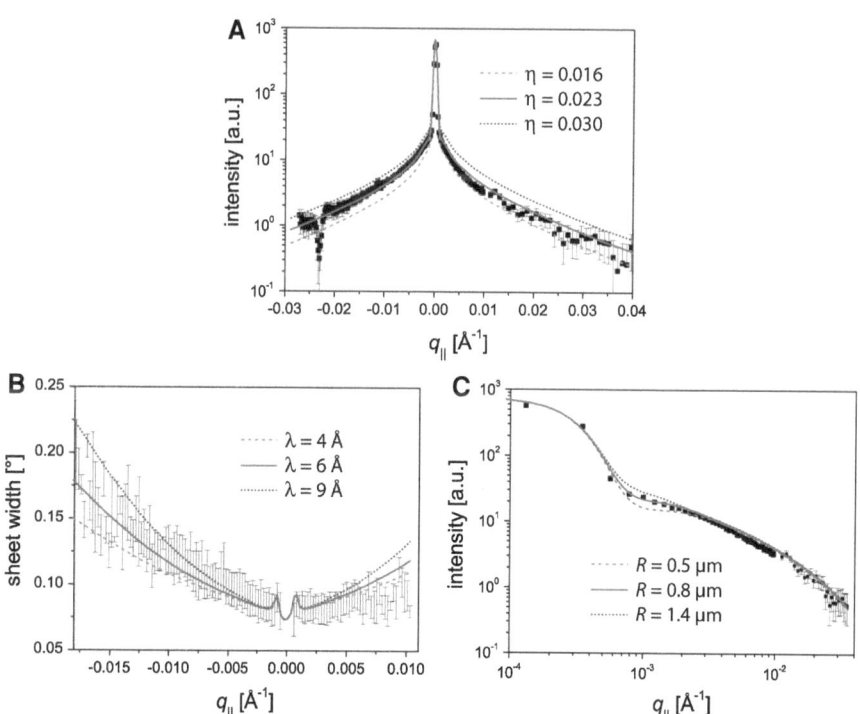

Fig. 4.5 Comparison of experimental data and simulation subject to variations of the model parameters. Slight deviations of each of the parameters result in a significant mismatch (characteristic for the parameter) between simulation and experimental data. System: Gent-D in H_2O atmosphere at $T = 80$ °C and $h_{rel} \approx 95\%$ (see Sect. 5.1.3). (**a**) Variation of the Caillé parameter η. (**b**) Variation of the de Gennes parameter λ. (**c**) Variation of the cut-off radius R

4.1.2.1 Beyond the Kinematic Approximation

To expand the range, where experimental data and simulations can be compared, the scattering intensities could be simulated in the dynamical DWBA in principle (see Sect. 2.2.3). However, detailed knowledge about the mesoscopic sample structure would have to be available. This includes the number of membrane layers N and their lateral distribution into domains, as well as a detailed model for the scattering length density profile of the membranes at each measurement condition. Moreover, DWBA simulations are numerically more costly by at least an order of magnitude, and the optimization of the free parameters requires an extremely high effort. In contrast, the approach taken in this thesis using kinematic approximation and the cut-off radius R is very robust and gives reliable results for a wide range of sample types as demonstrated in Chaps. 5 and 6.

4.1.2.2 Multiple Scattering

Multiple scattering, which denotes the scattering of an already scattered beam, was observed previously for multilayers of hard-matter [3, 6]. This phenomenon is also observed for the (soft-matter) membrane multilayers studied in this work (see Sect. 5.2.3 and Fig. 5.13). Multiple scattering can be quantitatively accounted for using DWBA. For a more intuitive understanding of this process, the positions of the double-scattering peaks observed in experiments are geometrically derived in the following.

Let the angle of incidence Ω, at which a beam impinges to a set of oriented rough multilayers, coincide with the first order Bragg angle $\theta_B = \arcsin(\lambda/2d)$. This scenario is depicted in Fig. 4.6 (top). The specularly reflected beam with the intensity $S(\Gamma = 2\theta_B, \Omega = \theta_B)$ may be scattered for a second time in a diffuse manner. The angle of incidence for this second process is $-\theta_B$, as determined by the preceding specular Bragg reflection. Under this condition the scattering function $S(\Gamma, \Omega = -\theta_B)$ possesses an intensity peak for $\Gamma = 2\theta_B$, and therefore the intensity of this secondary scattering process is proportional to $S(\Omega = -\theta_B, \Gamma = 2\theta_B)$. The resulting scattering angle (i.e., the detection angle Γ) for this double scattering process is $\Gamma = 4\theta_B$, as shown in the figure. Thus, a peak in the double scattering intensity I_{DS} is found for $\Omega = \theta_B$ and $\Gamma = 4\theta_B$, with:

$$I_{DS}(\Omega = \theta_B, \Gamma = 4\theta_B) \propto S(\Omega = \theta_B, \Gamma = 2\theta_B) \cdot S(\Omega = -\theta_B, \Gamma = 2\theta_B)$$

This peak represents a specular first order Bragg reflection followed by a diffuse first order Bragg reflection. An analogous double scattering process is depicted in the same figure (bottom) and represents the reverse sequence of single scattering processes, with the intensity:

$$I_{DS}(\Omega = 3\theta_B, \Gamma = 4\theta_B) \propto S(\Omega = 3\theta_B, \Gamma = 2\theta_B) \cdot S(\Omega = \theta_B, \Gamma = 2\theta_B)$$

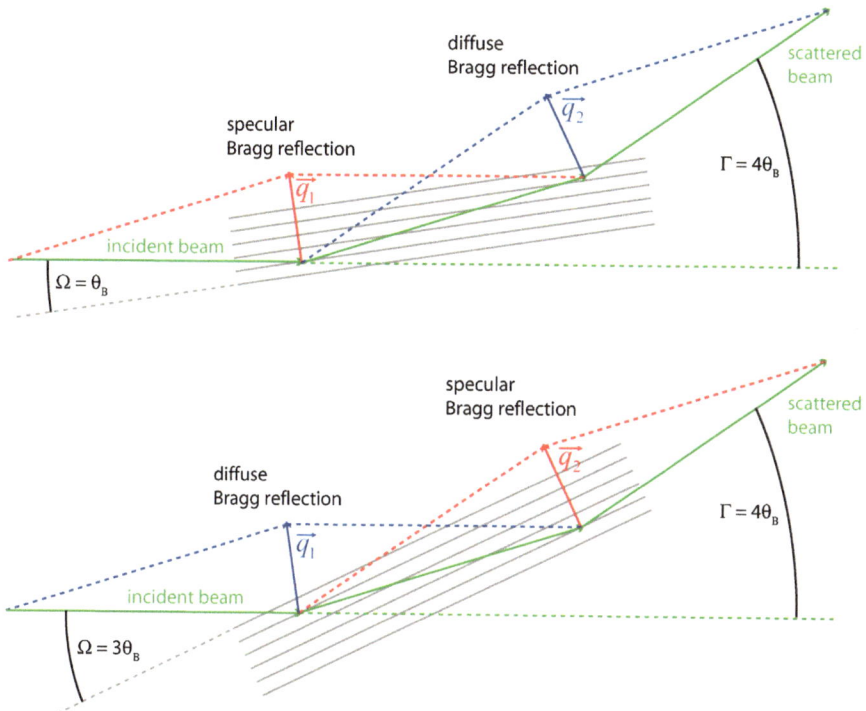

Fig. 4.6 Geometrical description of double scattering in multilayer systems. θ_B denotes the Bragg angle of the multilayers. (*Top*) specular Bragg reflection ($\vec{q_1}$) followed by diffuse Bragg reflection ($\vec{q_2}$). A peak is observed at $\Omega = \theta_B$ and $\Gamma = 4\theta_B$. (*Bottom*) diffuse Bragg reflection ($\vec{q_1}$) followed by specular Bragg reflection ($\vec{q_2}$). A peak is observed at $\Omega = 3\theta_B$ and $\Gamma = 4\theta_B$

For reason of symmetry:

$$I_{DS}(\Omega = \theta_B, \quad \Gamma = 4\theta_B) = I_{DS}(\Omega = 3\theta_B, \quad \Gamma = 4\theta_B)$$

The corresponding double scattering peaks are found at the following reciprocal space coordinates (see Fig. 4.7):

$$q_z = \frac{2\pi}{\lambda}(\sin\theta_B + \sin3\theta_B) \cong \frac{4\pi}{d}, \quad q_{\|} = \pm\frac{2\pi}{\lambda}(\cos\theta_B - \cos3\theta_B) \cong \pm\frac{2\pi}{\lambda}\left(\frac{\lambda}{d}\right)^2$$

These peaks possess a high intensity only if sufficiently strong out-of-plane multilayer ordering gives rise to a strong first order specular reflection and at the same time the topological roughness of the multilayers is sufficiently high to give rise to strong diffuse Bragg sheets (see Sect. 5.2.3).

Fig. 4.7 Position of the double scattering peaks in the reciprocal space. *Blue* and *red arrows* coincide with those depicted in Fig. 4.6

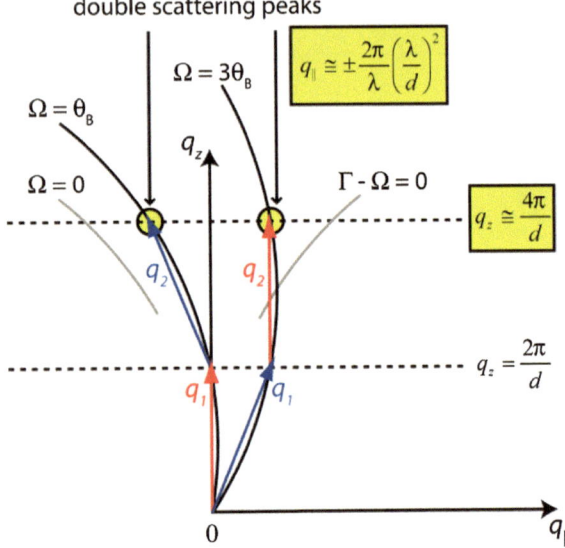

4.1.3 Summary of Sect. 4.1

A method for the determination of the mechanical properties of interacting membranes from specular and off-specular neutron scattering was developed. This required the calculation of membrane displacement correlation functions based on the compression modulus B and the membrane bending rigidity κ. To generate realistic sets of correlation functions, a "cut-off radius" was introduced as a free parameter that successfully accounts for the mesoscopic sample structure in a phenomenological manner. In order to theoretically model the experimentally obtained two-dimensional scattering intensity datasets using membrane displacement correlation functions, the kinematic description of scattering from stratified interfaces was theoretically adapted for the case of periodic membrane multilayers. The limitations of the kinematic approximation were explored and scattering features not captured by this approximation, such as double scattering peaks, were derived from geometrical considerations. The presented method enables the comprehensive and accurate description of experimentally obtained two-dimensional scattering intensity datasets by the underlying continuum mechanical model, and thus the reliable determination of mechanical parameters of interacting membranes (see Chaps. 5 and 6).

4.2 Electrostatic Interactions between Charged Lipid Membranes

In the experimental results part of this thesis, the interaction of charged phospholipid membranes in various buffers is quantitatively modeled (see Sect. 5.2).

This requires the accurate consideration of electrostatic interactions in mixed electrolytes (i.e., cations and anions of different valences dissolved in water). Since there are no general algebraic solutions to this problem, the numerical approach taken in this thesis is developed in the following.

In a continuum approximation the charged surface of a lipid membrane can be considered as a plane surface with a homogenous charge density (see Sect. 2.1.2). The electrostatic interaction of such a surface with its environment strongly depends on the ion concentration of the aqueous environment. This is discussed for a single charged surface and for two interacting charged surfaces, representing two charged lipid membranes.

4.2.1 A Single Charged Surface in an Electrolyte

According to Grahame [7], the electric potential $\psi(x)$ at distance x from a charged surface in an electrolyte is described by the differential equation

$$\frac{d\psi}{dx} = \left(2\varepsilon\varepsilon_0 k_B T \sum_m \rho_{0m}[\exp(-z_m e\psi/k_B T) - 1] \right)^{\frac{1}{2}}, \tag{4.4}$$

with the boundary condition

$$\psi(x = 0) = \psi_0.$$

Here, $k_B T$ denotes the thermal energy, ε_0 (ε) the (relative) dielectric constant, ρ_{0m} and z_m the bulk density and charge number of ion species m, and e the elementary charge. The electric potential at the surface, ψ_0, is determined [8] by the surface charge density σ:

$$\sigma^2 = 2\varepsilon\varepsilon_0 k_B T \sum_m \rho_{0m}[\exp(-z_m e\psi_0/k_B T) - 1] \tag{4.5}$$

ψ_0 can be calculated by numerically inverting Eq. 4.5. Near the surface the ion density profiles follow the Boltzmann distribution.

$$\rho_m(x) = \rho_{0m}\exp(-z_m e\psi(x)/k_B T).$$

Algebraic solutions to Eq. 4.4 are only available for few simple types of electrolytes [7, 8], and not applicable to mixed electrolytes. Therefore, Eq. 4.4 was solved numerically with the Maple software package (Maplesoft) using a Runge–Kutta Fehlberg method [9] to obtain the x-dependence of the electric potential and the ion densities. Figure 4.8 (left panel) shows the electric potential calculated for $\sigma = -10$ mC/m^2. The electrolyte solution contains 30 mM monovalent chloride anions, 10 mM monovalent sodium cations, and 10 mM divalent calcium cations (corresponding to [NaCl] = [CaCl$_2$] = 10 mM). The right panel of the figure shows the resulting concentration profiles of Cl$^-$, Na$^+$, and Ca^{2+}.

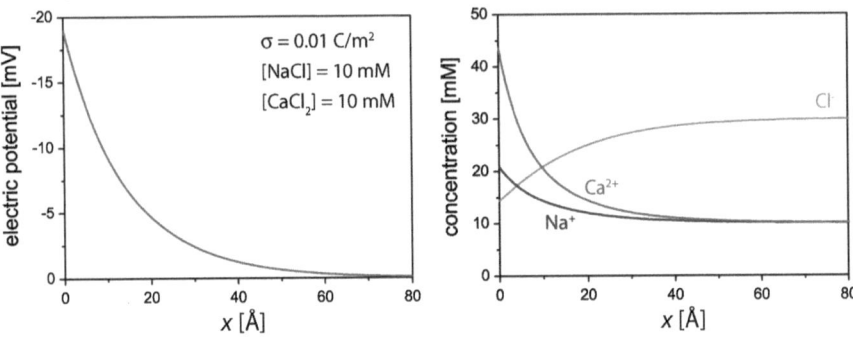

Fig. 4.8 Calculated electric potential (*left*) and concentrations of monovalent and divalent ions (*right*) near a single charged surface

An important feature of the ion density profiles is that the surface charge is fully compensated by the ion enrichment/depletion:

$$\sum_m z_m e \int_0^\infty (\rho_m - \rho_{0m})dx = -\sigma$$

In practice, this is valid if the upper integration boundary is large compared to the Debye screening length κ^{-1}.

$$\kappa^{-1} = \left(\sum_m \rho_{0m} e^2 z_m^2 / (\varepsilon \varepsilon_0 k_B T) \right)^{-1/2}.$$

As discussed in Sect. 2.1.2, the here presented treatment of screened electrostatics relies on a continuum assumption, which loses its validity at length scales comparable to the atomic structure of the media. Moreover, great care has to be taken while dealing with the combination of high charge densities and high salt concentrations, where the continuum description breaks down. Within limits, this problem can be solved by taking a layer of strongly bound ions, called Helmholtz or Stern layer, into account [8]. A more realistic description is provided by computer simulations that explicitly account for discrete ions [10–12] (see Chap. 6), at the cost of much higher numerical effort.

4.2.2 Two Charged Surfaces in an Electrolyte

The electric potential $\psi(x)$ between two surfaces with identical charge density in an electrolyte can be calculated by numerically integrating the Poisson-Boltzmann equation.

$$\frac{d^2\psi}{dx^2} = -\frac{\rho}{\varepsilon\varepsilon_0} = -\frac{1}{\varepsilon\varepsilon_0}\sum_m z_m e\rho_{0m}\exp(-z_m e\psi/k_B T). \tag{4.6}$$

The boundary conditions [13] depend on the separation of the surfaces, d_W.

1. $\left.\dfrac{d\psi}{dx}\right|_{x=0} = 0.$

2. $\left.\dfrac{d\psi}{dx}\right|_{x=d_W/2} = -\dfrac{\sigma}{\varepsilon\varepsilon_0}.$

Here, $x = 0$ coincides with the "midplane" (i.e., the center between the two surfaces). Boundary condition 1 accounts for the symmetry requirement, while boundary condition 2 accounts for the electric potential gradient at the surfaces, which is proportional to their charge density σ. This boundary value problem was solved numerically with the Maple software package (Maplesoft) using a finite difference technique with Richardson extrapolation [9]. Figure 4.9 shows the calculated electric potential $\psi(x)$ between the charged surfaces for two different surface separations d_W. Again, $\sigma = -10$ mC/m^2 and [NaCl] = [CaCl$_2$] = 10 mM was used for the calculations. Note that for the smaller surface separation ($d_W = 50$ Å), the "midplane potential" ψ_{mp} (i.e., the electric potential at $x = 0$) has a higher absolute value than for the larger separation ($d_W = 100$ Å). With increasing separations the midplane potential decays to zero. The right panel of the figure shows the corresponding ion concentration profiles of Cl$^-$, Na$^+$, and Ca^{2+} for $d_W = 50$ Å.

The repulsive pressure Π_{ES} resulting from the electrostatic interaction between two charged surfaces in an electrolyte is given by [8]:

$$\Pi_{ES} = k_B T \sum_i \rho_{0,i}\left(\exp\left(-z_i e\psi_{mp}/k_B T\right) - 1\right), \tag{4.7}$$

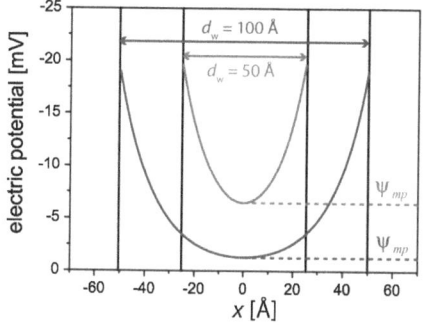
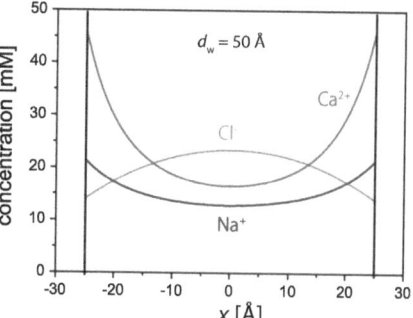

Fig. 4.9 (*Left*) calculated electric potential between two charged surfaces for two different surface separations d_W. The midplane potential ψ_{mp} depends on d_W. (*Right*) calculated concentrations of monovalent and divalent ions between two charged surfaces for $d_W = 50$ Å

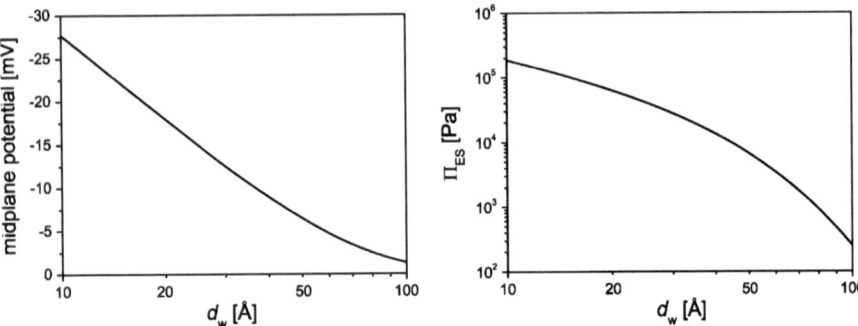

Fig. 4.10 Midplane potential ψ_{mp} (*left*) and repulsive electrostatic pressure Π_{ES} (*right*) as a function of d_W

meaning that the repulsion can be expressed with the osmotic pressure created by the ion enrichment at the midplane, which is determined by the midplane potential ψ_{mp}. To compute the dependence of the electrostatic repulsion on the surface separation, ψ_{mp} was calculated as a function of d_W, which required each time the integration of Eq. 4.6. Figure 4.10 (left panel) shows ψ_{mp} in a wide range of d_W for the above mentioned charge density and electrolyte. The right panel shows the corresponding repulsive pressure Π_{ES}. In both panels the d_W-axis starts at 10 Å, as the continuum approximation loses its validity for small surface separations (see Sect. 2.1.2).

4.2.3 The Weak-Overlap Approximation

The numerical effort for the calculation of $\Pi_{ES}(d_W)$ can be dramatically reduced if the electrostatic interaction of the surfaces is assumed to be weak. In this case the midplane potential can be approximated as twice the (undisturbed) electric potential ψ in a distance of $x = d_W/2$ form an isolated charged surface [8] (see above):

$$\psi_{mp}(d_W) \cong 2\psi(x = d_W/2).$$

While for the exact solution it is required to integrate Eq. 4.6 for each value of d_W, Eq. 4.4 only needs to be integrated once if the weak-overlap approximation is taken. Figure 4.11 shows a comparison of the approximation with the results of the exact calculation for $\sigma = -10$ mC/m^2 at various electrolyte conditions. Generally, the overlap is weak for surface separation much larger than the Debye screening length ($d_W \gg \kappa^{-1}$). However, the approximation appears surprisingly good throughout the probed range (10 Å $< d_W <$ 200 Å) at sufficiently high salt concentrations (here, [NaCl] = [CaCl$_2$] = 10 mM). In contrast, at low salt concentrations the approximation is seen to be rather poor for small surface separations. Throughout this thesis the exact calculation is used.

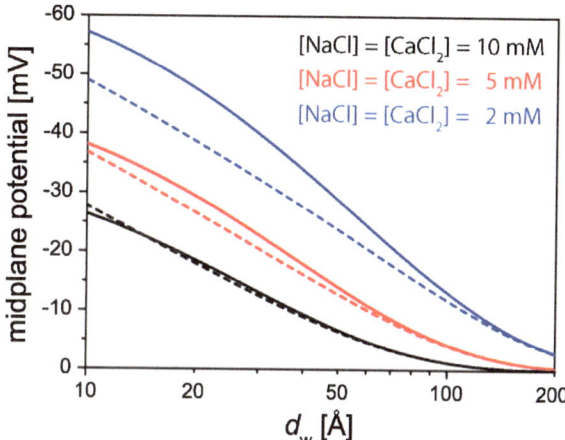

Fig. 4.11 Midplane potential ψ_{mp} in weak-overlap approximation (*solid lines*) and exact result (*dashed lines*) as a function of the surface separation d_W for various ionic strengths. For low salt concentrations, the approximation only holds at large separations

4.2.4 Summary of Sect. 4.2

A method for the numerical calculation of electric potentials and ion density distributions near a single charged surface and between two equally charged surfaces in mixed electrolytes was developed. It allows for the calculation of repulsive interactions between two surfaces as a function of their separation. Furthermore, the range of validity of the commonly used "weak-overlap" approximation was explored and discussed. The here presented method enables the quantitative description of the electrostatic repulsion between charged membranes in mixed electrolytes (see Sect. 5.2).

4.3 Interpretation of X-Ray Fluorescence Signals

In the experimental results part of this thesis, a recently introduced technique called grazing-incidence X-ray fluorescence (GIXF) is used. The interpretation of the fluorescence signals requires a considerable mathematical effort. However, despite quite a few articles on GIXF [14–18] literature lacks a comprehensive description of this new technique. In the following a method for the interpretation of X-ray fluorescence signals is developed and practical aspects are discussed. The experimental setup is illustrated in the Sect. 3.3.1.3 (Fig. 3.9).

The fluorescence intensity emitted in an GIXF experiment at the air/water interface by an ion species of interest depends on the angle of incidence of the X-ray beam. This dependency is given as follows [18]:

$$I_m^{\text{fluo}}(q_z) = \int_0^\infty I_{q_z}^{illu}(z) \cdot c_m(z) \cdot \exp(-\mu_m z)dz \qquad (4.8)$$

Here, $q_z = \frac{4\pi}{\lambda}\sin\alpha_i$ (see Sect. 3.3.1.3), m indicates the ion species (or, more generally, a chemical element), $c_m(z)$ denotes its vertical concentration profile, μ_m the absorption coefficient of the bulk buffer for the fluorescence radiation emitted by species m, and $I_{q_z}^{illu}(z)$ the vertical intensity profile of the incident illumination, which depends on the electronic structure of the monolayer. Once this illumination profile is known, the ion concentration profiles can be reconstructed from the measured fluorescence intensities. The electronic structure of a monolayer at the air/water interface can be parameterized using slab models, where each slab represents a layer of constant electron density within the monolayer (see Sect. 2.2.1.2). For low q_z (not much larger than q_z^c) the interfacial roughness between the slabs can be neglected, as the Névot-Croce factors, $\exp(-q_z^2\sigma^2)$, are close to unity (see Sect. 2.2.2.2).

4.3.1 Calculation of Illumination Profiles

The calculation of the illumination intensity profile $I_{q_z}^{illu}(z)$ is non-trivial and is explained in the following. An illustration is given in Fig. 4.12. A good starting point for this type of calculations is the book of Born and Wolf [19].

Be E_0 the amplitude at $z = 0$ of an electromagnetic wave impinging from medium 0 to a set of $N - 1$ stratified slabs. Media 0 and N are semi-infinite bulk media. The z-components (perpendicular to the interfaces) of the wave vectors in each medium, k_j, depend on q_z and on the refractive indices n_j of the media, and thus on the electron densities ρ_j (see Sect. 2.2.2):

$$k_j^z = \frac{2\pi}{\lambda}\sqrt{\left(\frac{q_z\lambda}{4\pi}\right)^2 + n_j^2 - 1}$$

In each medium j ranging from z_j to z_{j+1}, the vertical intensity profile is the absolute square of the sum of the electromagnetic waves E_+ and E_-, propagating in positive and negative z directions, respectively.

$$I_{q_z}^{illu}(z) = |E_+(q_z, z) + E_-(q_z, z)|^2$$

In the first medium ($j = 0$):

$$E_+^0(q_z, z) = E_0\, e^{ik_0^z(z-z_1)}$$

$$E_-^0(q_z, z) = E_0\, r_{0,N} \cdot e^{ik_0^z(z_1-z)}$$

$$I_{q_z}^{illu}(z < z_1) = \left|\left(e^{ik_0^z(z-z_1)} + r_{0,N}\, e^{ik_0^z(z_1-z)}\right) \cdot E_0\right|^2$$

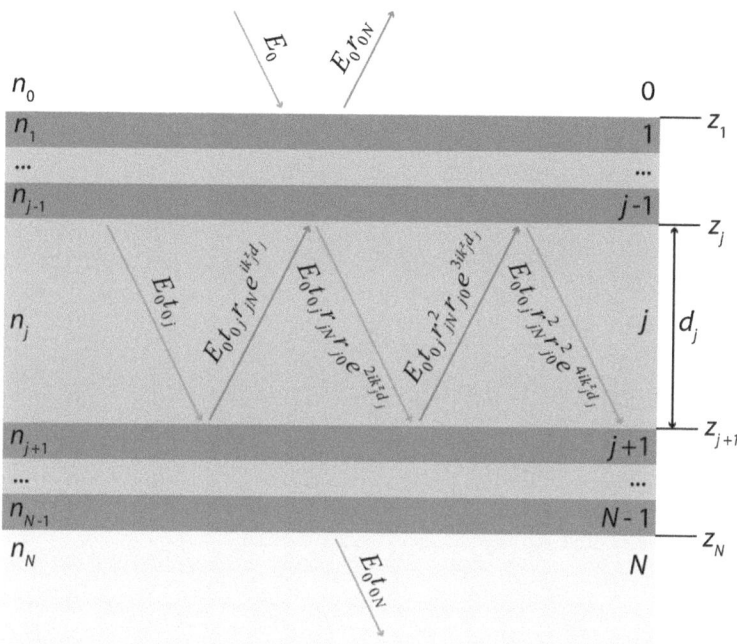

Fig. 4.12 Sketch of the electromagnetic waves propagating in positive (*green arrows*) and negative (*red arrows*) z-directions in a stratified system of homogenous slabs that is illuminated with the incident wave E_0

$E_+^0(q_z, z)$ coincides with the amplitude of the incident wave, while $E_-^0(q_z, z)$ represents the amplitude of the wave reflected by the layered system and $r_{0, N}$ denotes the Parratt amplitude reflection coefficient (see Sect. 2.2.2.3) for a beam traveling from medium 0 to medium N.

In the last medium ($j = N$):

$$E_+^N(q_z, z) = E_0 \, t_{0, N} \cdot e^{ik_N^z(z - z_N)}$$

$$E_-^N(q_z, z) = 0$$

$$I_{q_z}^{illu}(z \geq z_N) = \left| E_0 \, t_{0, N} \, e^{ik_N^z(z - z_N)} \right|^2$$

$E_+^N(q_z, z)$ represents the amplitude of the wave transmitted through the entire layered system and $t_{0, N}$ denotes the Parratt amplitude transmission coefficient (see Sect. 2.2.2.3) for a beam traveling from medium 0 to medium N. $E_-^N(q_z, z)$ vanishes, as there is no more reflected wave in the last medium.

Within the slabs ($1 \leq j \leq N - 1$) multiple reflections have to be considered. This is illustrated in Fig. 4.12, and requires an infinite summation to obtain $E_+^j(q_z, z)$ and $E_+^j(q_z, z)$:

$$E_+^j(q_z,z) = E_0 \cdot \left(t_{0,j} + t_{0,j}\, r_{j,N}\, r_{j,0}\, e^{2ik_j^z d_j} + t_{0,j}\, r_{j,N}^2\, r_{j,0}^2\, e^{4ik_j^z d_j} + \ldots \right) \cdot e^{ik_j^z(z-z_j)}$$

$$= E_0\, t_{0,j} \cdot \left(1 + r_{j,N}\, r_{j,0}\, e^{2ik_j^z d_j} + \left(r_{j,N}\, r_{j,0}\, e^{2ik_j^z d_j} \right)^2 + \ldots \right) \cdot e^{ik_j^z(z-z_j)}$$

$$= E_0\, t_{0,j} \cdot e^{ik_j^z(z-z_j)} \cdot \sum_{l=0}^{\infty} \left(r_{j,N}\, r_{j,0}\, e^{2ik_j^z d_j} \right)^l$$

$$= \frac{t_{0,j}}{1 - r_{j,N}\, r_{j,0}\, e^{2ik_j^z d_j}} \cdot E_0 \cdot e^{ik_j^z(z-z_j)}$$

And similarly:

$$E_-^j(q_z,z) = E_0 \cdot \left(t_{0,j}\, r_{j,N}\, e^{ik_j^z d_j} + t_{0,j}\, r_{j,N}^2\, r_{j,0}\, e^{3ik_j^z d_j} + t_{0,j}\, r_{j,N}^3\, r_{j,0}^2\, e^{5ik_j^z d_j} + \ldots \right) \cdot e^{ik_j^z(z_{j+1}-z)}$$

$$= \frac{t_{0,j}\, r_{j,N}\, e^{ik_j^z d_j}}{1 - r_{j,N}\, r_{j,0}\, e^{2ik_j^z d_j}} \cdot E_0 \cdot e^{ik_j^z(z_{j+1}-z)}$$

Finally:

$$I_{q_z}^{illu}(z_j \leq z < z_{j+1}) = \left| \frac{t_{0,j}}{1 - r_{j,N}\, r_{j,0}\, e^{2ik_j^z d_j}} \left(e^{ik_j^z(z-z_j)} + r_{j,N}\, e^{ik_j^z d_j}\, e^{ik_j^z(z_{j+1}-z)} \right) \cdot E_0 \right|^2$$

Here, $d_j = z_{j+1} - z_j$ denotes the thickness of the jth medium (see Fig. 4.12), and $r_{x,y}$ and $t_{x,y}$ the Parratt amplitude reflection and transmission coefficients for a beam traveling from medium x to medium y. In practice, this means that an $N \times N$ matrix of Parratt coefficients has to be calculated before calculating an illumination intensity profile. For a beam traveling in positive z-direction the commonly used recursion formula for the reflection coefficients can be used (see Sect. 2.2.2.3), together with the corresponding formulae for the transmission coefficients [20, 21].

$$r_{j,N} = \frac{r_{j,j+1}^F + r_{j+1,N} \cdot e^{2ik_{j+1}^z d_{j+1}}}{1 + r_{j,j+1}^F \cdot r_{j+1,N} \cdot e^{2ik_{j+1}^z d_{j+1}}}, \qquad r_{N-1,N} = r_{N-1,N}^F$$

$$t_{j,N} = \frac{t_{j,j+1}^F \cdot t_{j+1,N} \cdot e^{ik_{j+1}^z d_{j+1}}}{1 + r_{j,j+1}^F \cdot r_{j+1,N} \cdot e^{2ik_{j+1}^z d_{j+1}}}, \qquad t_{N-1,N} = t_{N-1,N}^F.$$

For a beam traveling in negative z-direction the inverse problem has to be solved:

$$r_{j,0} = \frac{r_{j,j-1}^F + r_{j-1,0} \cdot e^{2ik_{j-1}^z d_{j-1}}}{1 + r_{j,j-1}^F \cdot r_{j-1,0} \cdot e^{2ik_{j-1}^z d_{j-1}}}, \qquad r_{1,0} = r_{1,0}^F$$

$$t_{j,0} = \frac{t^F_{j,j-1} \cdot t_{j-1,0} \cdot e^{ik^z_{j-1}d_{j-1}}}{1 + r^F_{j,j-1} \cdot r_{j-1,0} \cdot e^{2ik^z_{j-1}d_{j-1}}}, \quad t_{1,0} = t^F_{1,0}.$$

Similarly, Ohta and Ishida [22] have presented a method to calculate the illumination profile intensity in stratified media for a light beam using a matrix formalism.

4.3.2 Calculation of X-Ray Fluorescence Intensities

In the following, X-ray fluorescence signals are modeled for two example systems that are closely related to the experimental system discussed in Sect. 6.3.

1. The blank interface between air and an aqueous buffer with a homogenous ion concentration c_0. This is illustrated in Fig. 4.13 (left).
2. A thin film at the interface between air and an aqueous buffer with bulk ion concentration c_0. This is illustrated in Fig. 4.13 (right).

The film is composed of two slab media, one of which (slab 1) has a thickness of 20 Å and is hydrophobic, while the other one (slab 2) has a thickness of 30 Å and is hydrophilic. X-ray absorption in both thin slabs is neglected ($\beta_1 = \beta_2 = 0$) and slab 1 possesses an electron density similar to that of water, while the electron density of slab 2 is about 50 % higher. Further, there are no ions in slab 1, but ions are enriched in slab 2.

In the first step, the illumination intensity profiles are calculated. The results are shown in Fig. 4.14, where $I^{illu}_{q_z}$ is plotted as a function of z for both systems at three different q_z values ($q_z = 0.70q^c_z$: green line, $q_z = 0.99q^c_z$: blue line, $q_z = 1.03q^c_z$: red line). Below q^c_z (green and blue lines) the intensity decays inside the aqueous buffer and only the vicinity of the interface is illuminated by an evanescent field (see Sect. 2.2.2). In contrast, above q^c_z (red line) the intensity propagates into the

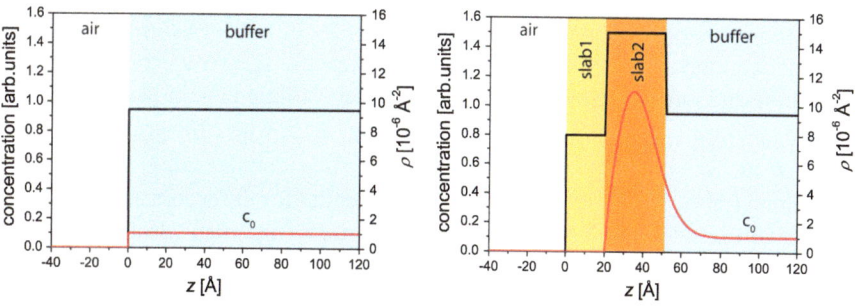

Fig. 4.13 Example slab models (*black lines*) and ion concentration profiles (*red lines*) of a blank buffer surface (*left*) and a monolayer film deposited at the air/buffer interface (*right*). The monolayer is composed of two slabs in one of which the ion concentration has a peak

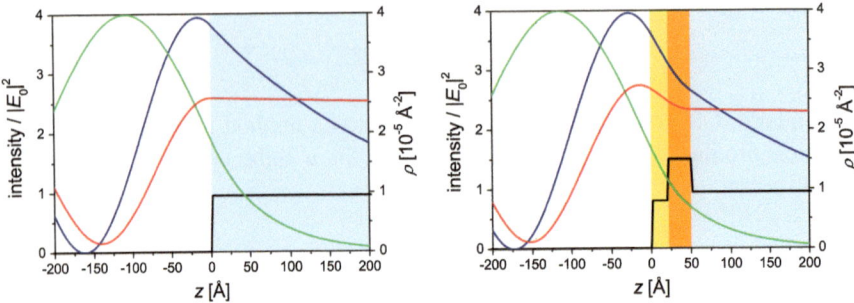

Fig. 4.14 X-ray illumination profiles $I_{q_z}^{illu}(z)$ for various q_z values ($q_z = 0.70q_z^c$: *green line*, $q_z = 0.99q_z^c$: *blue line*, $q_z = 1.03q_z^c$: *red line*). (*Left*) blank buffer surface. (*Right*) monolayer film deposited at the air/buffer interface

Fig. 4.15 X-ray illumination profile deformed at according to the electronic structure of the monolayer film (*solid line*) and corresponding illumination profile near the blank buffer surface (*dashed line*)

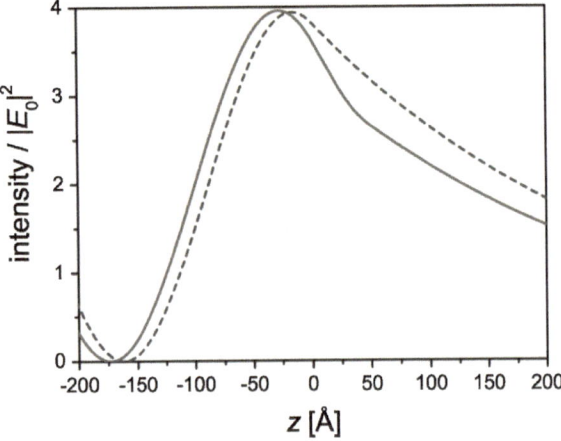

bulk and the beam is only damped due to the absorption by the bulk medium (here, $\beta_b = 10^{-8}$).

The comparison between the illumination profiles of both systems (without and with the monolayer film) yields significant differences. These reflect the influence (refraction and interference) of the electron density variation due to presence of the film. The deformation of $I_{q_z}^{illu}$ is highlighted in Fig. 4.15, where the profiles for both systems at $q_z = 0.99q_z^c$ are superimposed. This demonstrates that correct accounting for the electronic film structure is essential for the accurate modeling of X-ray fluorescence signals. This becomes even more important for thicker films.

In the next step, fluorescence signals are computed for both systems by evaluating Eq. 4.8 as a function of q_z. This is numerically very costly, as a high z-resolution (at the Å scale) is required while at the same time the integration has to cover a large z-range (up to the 100 μm scale) before absorption has sufficiently damped the integrand (see Eq. 4.8). The resulting signals are shown in Fig. 4.16.

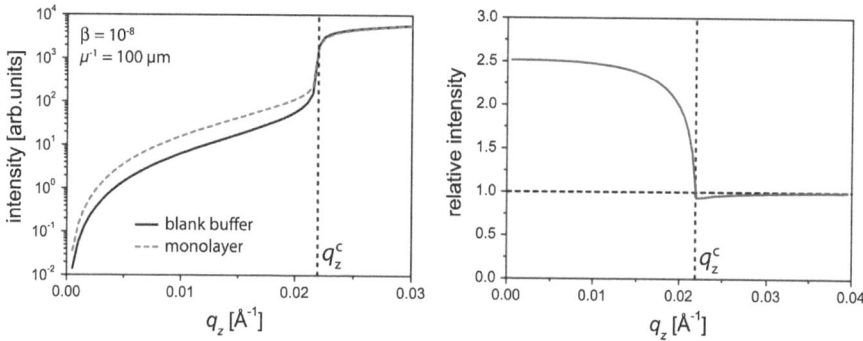

Fig. 4.16 (*Left*) calculated fluorescence intensities without (*solid line*) and with (*dashed line*) monolayer. (*Right*) "Buffer-normalized" fluorescence intensity of the monolayer system. *Vertical dashed lines* indicate the position of $q_z^c \cong 0.022\ \text{Å}^{-1}$

For the calculations $\mu^{-1} = 100$ μm and $\beta_b = 10^{-8}$ was used, where β_b represents the absorption of the aqueous buffer for the illuminating X-ray beam and μ^{-1} the decay length of the fluorescence radiation in the buffer. Both curves increase monotonically with q_z and show a massive increase around q_z^c. Below q_z^c, the signal in the presence of the film is significantly higher than that of the blank interface. This difference can be highlighted by dividing the film signal by the blank interface signal (Fig. 4.16, right panel), and reflects the enrichment of ions in slab 2 (Fig. 4.13). This representation will be referred to as "buffer-normalized" intensity in the following.

4.3.2.1 Absorption and Buffer-Normalized Fluorescence Intensities

The global shape of the fluorescence signal depends on the choice of the absorption parameters β_b and μ^{-1}. This is illustrated in Fig. 4.17, where modeled X-ray fluorescence signals from the blank interface and from the film are presented for various parameter sets. For weak absorption (here: $\beta_b = 10^{-10}$), the signals show a maximum (closely related to the Vineyard transmission function [23]) at $q_z = q_z^c$, and the signal intensities above q_z^c scale with μ^{-1}.

On the other hand, for stronger absorption (here: $\beta_b = 10^{-8}$), the signal intensities are largely independent of μ^{-1}, provided μ^{-1} is large enough. In fact, in the shown case the curves are practically identical as soon as μ^{-1} exceeds 20 μm (see Fig. 4.17 bottom right panel), and the re-absorption of fluorescence radiation (characterized by μ^{-1}) can be neglected. In practice, β_b depends on the photon energy used in the experiment, and various global shapes have been experimentally observed [16, 17].

As seen in Fig. 4.17, β_b and μ^{-1} influence the signal from the blank interface in the virtually identical way as they influence the signal from the film. As a consequence, the buffer-normalized fluorescence intensities are essentially identical for all sets of parameters β_b and μ^{-1}. This is shown in Fig. 4.18, where all

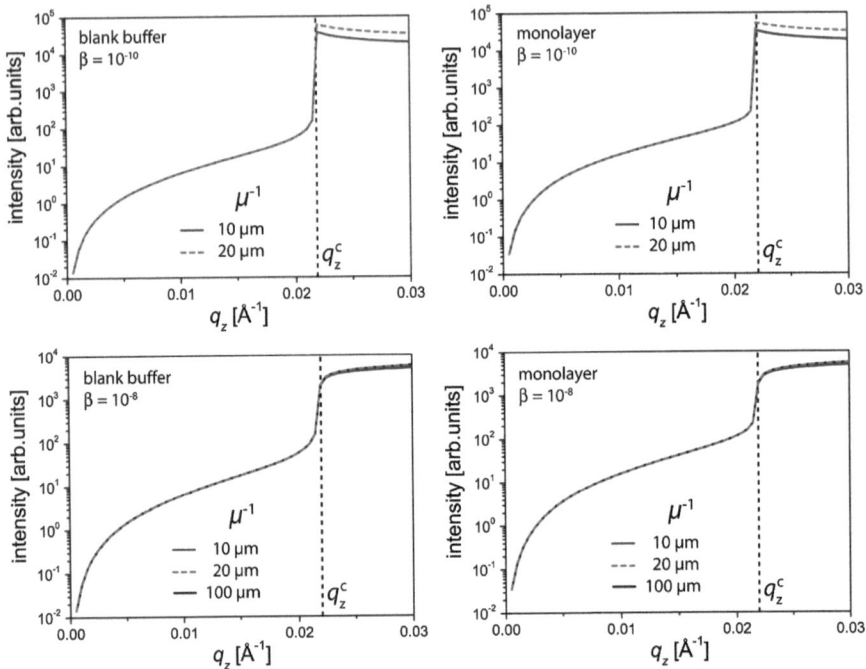

Fig. 4.17 X-ray fluorescence intensities calculated for various fluorescence radiation decay lengths μ^{-1} in case of weak (*top*, $\beta_b = 10^{-10}$) and strong (*bottom*, $\beta_b = 10^{-8}$) absorption of the illuminating X-ray beam by the bulk buffer. *Left* blank buffer surface. *Right* monolayer film

Fig. 4.18 Superimposed buffer normalized X-ray fluorescence intensities of the monolayer film calculated for various sets of absorption parameters (β_b and μ^{-1}, see Fig. 4.17). All the curves are virtually identical

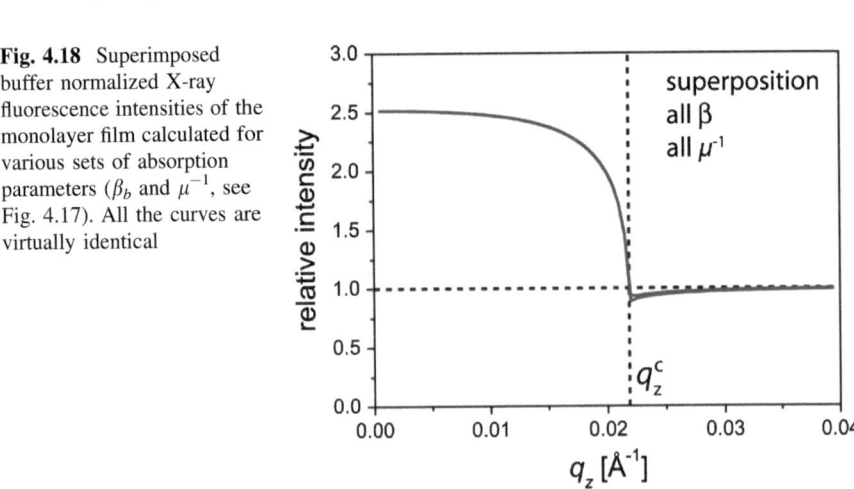

buffer-normalized intensities are superimposed and the curves are seen to be practically undistinguishable. This conclusion is valuable for the interpretation of GIXF signals: Although great care has to be taken while globally modeling the fluorescence signals from blank buffer and from the film individually, absorption aspects may be neglected while modeling the buffer normalized intensities. This makes the buffer-normalized fluorescence signals very reliable and resistant against possible absorption parameter uncertainties. Furthermore, other difficulties, such as the correct consideration of footprint size and fluorescence detector aperture, can be avoided in this way, as they again apply equally to the signal from blank buffer and from the film.

4.3.3 Summary of Sect. 4.3

A method for the interpretation of GIXF measurements with biological model systems at the air/water interface was developed. The calculation of X-ray illumination profiles from a slab model representation of the model systems was mathematically derived. X-ray fluorescence signals were modeled and discussed for two example systems that are closely related to the samples studied in Sect. 6.3: the blank air/buffer interface and a thin, layered film at the air/buffer interface. The implications of X-ray absorption were systematically investigated. It was concluded, that these issues can be avoided by interpreting "buffer-normalized" X-ray fluorescence intensities of the film. The presented method enables the accurate determination of ion density profiles near biological model systems at the air/water interface (see Sect. 6.3).

References

1. N. Lei, C.R. Safinya, R.F. Bruinsma, Discrete Harmonic Model for stacked membranes—theory and experiment. J. Phys. II **5**, 1155 (1995)
2. E.A.L. Mol, J.D. Shindler, A.N. Shalaginov, W.H. de Jeu, Correlations in the thermal fluctuations of free-standing smectic-A films as measured by X-ray scattering. Phys. Rev. E **54**, 536 (1996)
3. S.K. Sinha, X-Ray diffuse-scattering as a probe for thin-film and interface structure. J. Phys. III **4**, 1543 (1994)
4. T. Salditt, Thermal fluctuations and stability of solid-supported lipid membranes. J. Phys. Condens. Matter **17**, R287 (2005)
5. T. Salditt, M. Vogel, W. Fenzl, Thermal fluctuations and positional correlations in oriented lipid membranes. Phys. Rev. Lett. **90**, 178101 (2003)
6. V. Holý, T. Baumbach, Nonspecular X-ray reflection from rough multilayers. Phys. Rev. B **49**, 10668 (1994)
7. D.C. Grahame, Diffuse double layer theory for electrolytes of unsymmetrical valence types. J. Chem. Phys. **21**, 1054 (1953)
8. J.N. Israelachvili, *Intermolecular and surface forces* (Academic Press Inc., London, 1991)
9. W.H. Press, S.A. Teukolsky, W.T. Vetterling, B.P. Flannery, *Numerical recipes in C* (Cambridge University Press, Cambridge, 1992)
10. R.G. Oliveira et al., Physical mechanisms of bacterial survival revealed by combined grazing-incidence X-ray scattering and Monte Carlo simulation. C. R. Chimie **12**, 209 (2009)
11. E. Schneck, E. Papp-Szabo, B.E. Quinn, O.V. Konovalov, T.J. Beveridge, D.A. Pink, M. Tanaka, Calcium ions induce collapse of charged O-side chains of lipopolysaccharides from *Pseudomonas aeruginosa*. J. R. Soc. Interface **6**, S671 (2009)

12. R.G. Oliveira et al., Crucial Roles of Charged Saccharide Moieties in Survival of Gram Negative Bacteria Revealed by Combination of Grazing Incidence X-ray Structural Characterizations and Monte Carlo Simulations. Phys. Rev. E **81**, 041901 (2010)

13. B.W. Ninham, V.A. Parsegian, Electrostatic potential between surfaces bearing ionizable groups in ionic equilibrium with physiologic saline solution. J. Theor. Biol. **31**, 405 (1971)

14. W. Bu, D. Vaknin, X-ray fluorescence spectroscopy from ions at charged vapor/water interfaces. J. Appl. Phys. **105**, 084911 (2009)

15. N.N. Novikova et al., X-ray fluorescence methods for investigations of lipid/protein membrane models. J. Synchrotron Rad. **12**, 511 (2005)

16. N.N. Novikova et al., Total reflection X-ray fluorescence study of Langmuir monolayers on water surface. J. Appl. Cryst. **36**, 727 (2003)

17. V. Padmanabhan, J. Daillant, L. Belloni, Specific ion adsorption and short-range interactions at the air aqueous solution interface. Phys. Rev. Lett. **99**, 086105 (2007)

18. W.B. Yun, J.M. Bloch, X-ray near total external fluorescence method: experiment and analysis. J. Appl. Phys. **68**, 1421 (1990)

19. M. Born, E. Wolf, *Principles of optics*, 6th edn. (Pergamon, New York, 1980)

20. L.G. Parratt, Surface studies of solids by total reflection of X-rays. Phys. Rev. **95**, 359 (1954)

21. F. Abelès, RECHERCHES THÉORIQUES SUR LES PROPRIÉTÉS OPTIQUES DES LAMES MINCES. J. Phys. Rad. **11**, 307 (1950)

22. K. Ohta, H. Ishida, Matrix formalism for calculation of the light beam intensity in stratified multilayered films, and its use in the analysis of emission spectra. Appl. Opt. **29**, 2466 (1990)

23. G.H. Vineyard, Grazing-incidence diffraction and the distorted-wave approximation for the study of surfaces. Phys. Rev. B **26**, 4146 (1982)

Chapter 5
Inter-Membrane Interactions and Mechanical Properties of Membranes Composed of Synthetic Glycolipids

In this chapter, well-defined model systems prepared from synthetic glycolipids with known structures are studied using specular and off-specular neutron scattering in order to quantify the influence of generic and specific saccharide-saccharide interactions on the structure and mechanics of membrane multilayers.

5.1 Influence of Molecular Structure: Cylindrical and Bent Saccharides

In the first step, two types of glycolipids with a distinct difference in their disaccharide headgroup are compared to investigate the influence of the chemical structure on the generic saccharide-saccharide interactions. Solid-supported multilayers (see Sect. 2.1.4.4) of synthetic glycolipid membranes, prepared from 100% gentiobiose lipid or 100% Lac1 lipid (see Sects. 3.1.1 and 3.2.2), were studied by specular and off-specular neutron scattering to examine the influence of the saccharide head group structure on their structural ordering and mechanical properties. The gentiobiose lipid possesses a "bent" disaccharide headgroup, while that of the Lac1 lipid is "cylindrical". In order to optimize the neutron scattering length density contrast, both chain-hydrogenated (Gent-H and Lac1-H) and chain-deuterated (Gent-D and Lac1-D) glycolipids were used for experiments. The samples were kept either in a temperature-controlled climate chamber for experiments under defined osmotic pressures, or in a liquid cell for experiments in bulk water. The model system is schematically illustrated in Fig. 5.1 (left). The right panel of the figure shows a reciprocal space map (i.e., scattering intensity as a function of the reciprocal space coordinates, see Sect. 4.1.2) recorded from gentiobiose lipid membrane multilayers at 80 °C and high relative humidity ($\approx 95\%$).

E. Schneck, *Generic and Specific Roles of Saccharides at Cell and Bacteria Surfaces*, Springer Theses, DOI: 10.1007/978-3-642-15450-8_5,
© Springer-Verlag Berlin Heidelberg 2011

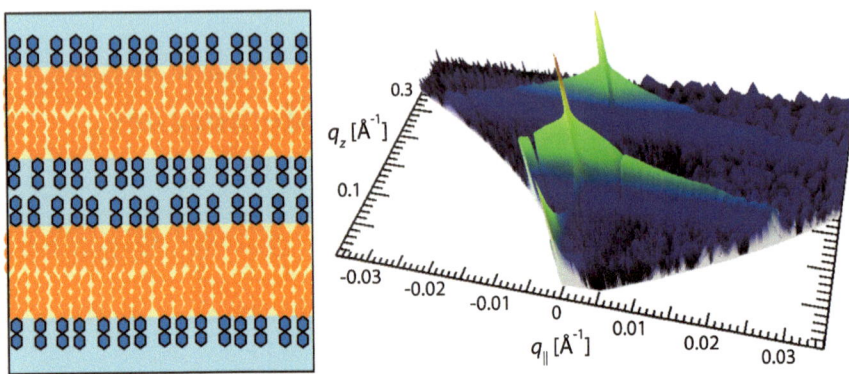

Fig. 5.1 Sketch of interacting glycolipid membrane membranes. Saccharide units are indicated with hexagons (*left*). Reciprocal space map recorded by specular and off-specular neutron scattering from gentiobiose lipid at 80 °C and high relative humidity ($\approx 95\%$) (*right*)

5.1.1 Phase Transitions of Glycolipid Membranes

The comprehensive investigation of structure and mechanics of glycolipid membrane multilayers requires knowledge on the phase behavior of the studied membrane systems. Only in this way can it be ensured, that the membranes assume the biologically relevant fluid L_α-phase throughout all measurement conditions.

Prior to this thesis, the thermotropic phase behavior (see Sect. 2.1.1.2) of gentiobiose lipid and Lac1 lipid suspended in excess amount of water were determined [1, 2] using small-angle X-ray scattering (SAXS) and differential scanning calorimetry (DSC). It was found that the phase transition temperature of these glycolipids strongly depends on length and conformation of the carbohydrate head groups: the gentiobiose lipid shows a thermotropic phase transition from gel (L_β) to fluid (L_α) phase at around $T_m = 43$ °C, while the Lac1 lipid undergoes a direct phase transition from crystalline (L_c) to fluid (L_α) phase (i.e., not mediated by L_β phase) at $T_m = 74$ °C. The "bent" gentiobiose headgroup of gentiobiose lipid shows no remarkable influence on the main phase transition temperature. In contrast, the phase transition enthalpy (30 kcal/mol) of Lac1 lipid is significantly higher than that of a phospholipid with the same chain length (15 kcal/mol for DPPC) [3]. This indicates that the "cylindrical" lactose headgroup has a significant influence on the phase transition enthalpy due to intra-membrane carbohydrate–carbohydrate interactions.

In this thesis the phase states of gentiobiose lipid and Lac1 lipid membranes were determined by neutron scattering under controlled humidity, to investigate the lyotropic phase behavior (i.e., the influence of the lipid/water ratio on the phase state) for a variety of temperatures, which enables the establishment of two-dimensional phase diagrams as a function of temperature and osmotic pressure (see Sect. 3.3.3.1).

The average length of a carbon bond in the hydrocarbon chains projected onto the normal to the membrane plane differs significantly [4] between the phase state of a membrane (see Sect. 2.1.1.2). This enables the identification of different phases from the lamellar periodicity of the membrane multilayers. The lamellar periodicity d of the stacks was calculated from the q_z positions of the Bragg peaks (see Eq. 2.2). Figure 5.2 shows the scattering intensities along the specular line ($q_{\parallel} \cong 0$) of solid-supported gentiobiose lipid membrane multilayers at 30 °C (top left) and 80 °C (top right), where the membranes take gel-like L_β- and fluid L_α-phase, respectively. In the vicinity of the melting transition temperature, phase coexistence can be observed. This is shown in the bottom left panel of the figure for gentiobiose lipid membrane multilayers at 55 °C and low relative humidity (corresponding to a high osmotic pressure ≈ 108 Pa). Here, the sets of Bragg peaks characteristic to both phase states are found. The phase diagram of gentiobiose lipid was determined (Fig. 5.2 bottom right panel) from the lamellar periodicities measured at various temperatures and osmotic pressures.

The transition temperature of the highly hydrated samples ($T_m \approx 45$ °C) obtained in this thesis is in good agreement with the corresponding values from

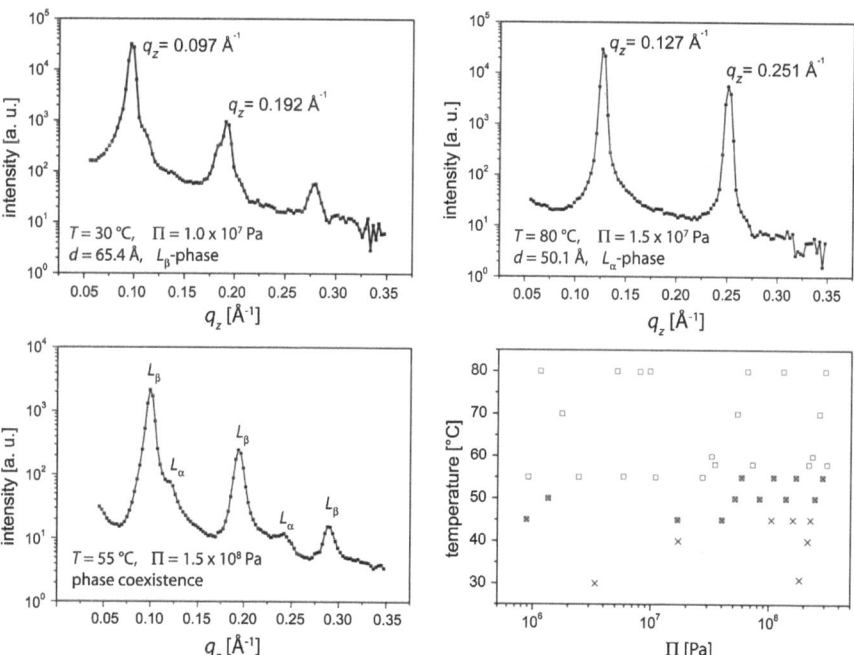

Fig. 5.2 Scattering intensities along the specular line ($q_z \cong 0$) of solid-supported gentiobiose lipid multilayers at 30 °C (*top left panel*) and at 80 °C (*top right panel*), where the membranes assume L_β- and L_α-phase, respectively. At 55 °C and low relative humidity (*bottom left panel*), the system exhibits phase coexistence. (*Bottom right panel*) Phase diagram of gentiobiose lipid membranes. L_α-phase is indicated by *squares*, L_β-phase by *crosses*, and phase coexistence by *both symbols*

DSC and X-ray measurements in the presence of excess water ($T_m = 43$ °C). In addition, at 55 °C, increasing the disjoining pressure over $\Pi \cong 4 \times 10^7$ Pa induces a lyotropic transition from L_α-phase to phase coexistence. Similarly, at 45 °C an increase of the disjoining pressure over $\Pi \cong >7 \times 10^7$ Pa is accompanied by a transition from phase coexistence to L_β-phase. Such pressure-induced phase transitions can be understood from a gain in chain ordering by exclusion of water molecules from the headgroup region.

Similarly, the phase diagram of Lac1 lipid was recorded but with less data points (not shown). The phase behavior of chain-hydrogenated and chain-deuterated molecules did not show significant differences. According to the high chain melting transition temperature of Lac1 lipid ($T_m = 74$ °C at full hydration), $T = 80$ °C was chosen for a systematic study of the influence of molecular structures on structural ordering and mechanical properties of the interacting membranes. At this temperature both Lac1 lipid and gentiobiose lipid are in fluid L_α-phase.

5.1.2 Modulation of Inter-Membrane Interactions via Saccharide Conformation

Due to the numerous relevant force contributions (see Sect. 2.1.2) a quantitative theoretical description of inter-membrane interactions is very difficult, especially for membranes composed of new types of molecules like gentiobiose lipid and Lac1 lipid. However, experimental force–distance relationships (see Sect. 2.1.2.7) enable the determination of quantitative relationships between the lamellar membrane periodicity d and the disjoining pressure Π (see Sect. 2.1.2.6) and can be interpreted to identify regimes of dominant force contributions. In this thesis, measurements of the lamellar periodicity at various osmotic pressures ($\Pi_{osm} = 5 \times 10^6$–$3 \times 10^8$ Pa, corresponding to $h_{rel} = 97$–15%) were conducted. The forces covered a high pressure regime ($\gtrsim 5 \times 10^7$ Pa) dominated by steric forces, typically modeled by a power law and a regime dominated by hydration forces, which are expected to follow an exponential decay at lower pressures ($\lesssim 5 \times 10^7$ Pa, see Sect. 2.1.2).

The force–distance relationships (Π vs. d) of gentiobiose lipid and Lac1 lipid multilayers at 80 °C are presented in a semi-logarithmic plot in Fig. 5.3. Here, the horizontal error bars correspond to the experimental errors of the periodicity measurement from the instrumental resolution, while the vertical ones coincide with the uncertainties of the relative humidity in the climate chamber. The solid lines show the decay models fitted to the data points within the two regimes (Note that, due to the limited range of the d-axis, the shape of the power law is almost undistinguishable from a straight line in the semi-logarithmic presentation.). In the high pressure regime, gentiobiose lipid membranes exhibit a characteristic exponent of $n = 13.9 \pm 1.7$, which is within the typical range to describe steric forces (see Sect. 2.1.2). In contrast, the value of Lac1 lipid membranes ($n = 7.0 \pm 1.7$) is

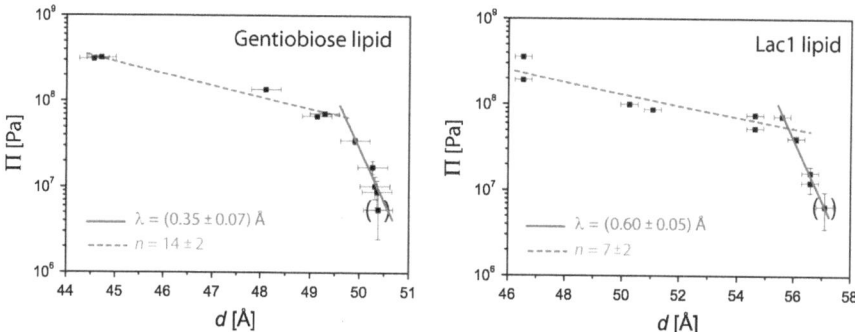

Fig. 5.3 Force–distance relationships of gentiobiose lipid (*left*) and Lac1 lipid (*right*), obtained from measurements of the lamellar periodicity at various relative humidities at 80 °C. Two different swelling regimes can be identified: steric repulsion at higher pressures ($\gtrsim 5 \times 10^7$ Pa) and hydration repulsion at lower pressures ($\lesssim 5 \times 10^7$ Pa). The *solid lines* correspond to the models fitted to the data points within the two pressure regimes

much smaller, indicating that the steric repulsion between the opposing membranes is weaker, possibly due to the flexibility of the head groups. This can be attributed to the difference in the finite compressibility of "cylindrical" lactose head groups and "bent" gentiobiose head groups in the direction normal to the membrane surface. At lower pressures, the characteristic hydration decay lengths obtained for both glycolipids [gentiobiose lipid: $\lambda_{\mathrm{hydr}} = (0.35 \pm 0.07)$ Å and Lac1 lipid: $\lambda_{\mathrm{hydr}} = (0.60 \pm 0.05)$ Å] are significantly lower than those reported for phosphatidylcholine lipid membranes ($\lambda_{\mathrm{hydr}} \cong 2.0$ Å, see Sect. 2.1.2.3). This indicates that neighboring glycolipid membranes are coupled more strongly than commonly studied phospholipid membranes in this pressure range, possibly due to additional attractive force contributions (such as "zipper" forces between carbohydrates) that compete with the hydration repulsion. Therefore, to gain insight into the force interplay at zero osmotic pressure, experiments in bulk water were conducted (see Sect. 5.1.3).

5.1.3   Influence of Saccharide Conformation on Membrane Mechanics

To reveal the influence of the saccharide structure on the mechanical properties of interacting glycolipid membranes, the off-specular scattering signals from solid-supported gentiobiose lipid and Lac1 lipid membrane multilayers were analyzed. For comparison, the same was done with membrane multilayers prepared from DPPC, a commonly studied phospholipid with the same type of all-saturated hexadecyl hydrocarbon chains (see Sect. 3.1.1). In order to optimize the scattering length density contrast, chain-deuterated DPPC (DPPC-D) was used. In the analysis of the scattering signals, the parameters η, λ, and R were varied to achieve

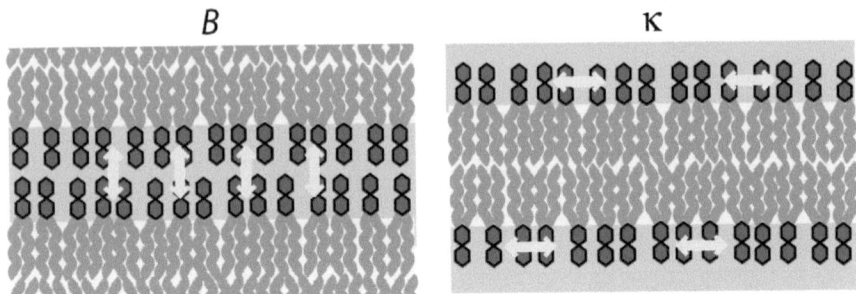

Fig. 5.4 Influence of inter-membrane saccharide interactions on the compression modulus *B* (*left*). Influence of the in-plane interactions of membrane-bound saccharides on the membrane bending rigidity *κ* (*right*)

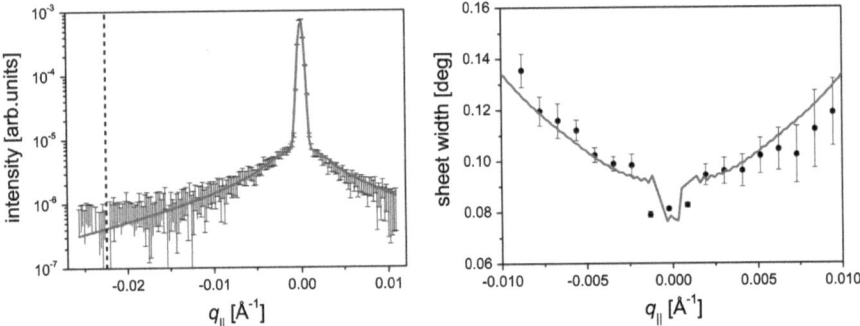

Fig. 5.5 Measured (*data points*) and simulated (*solid line*) second Bragg sheets of DPPC membrane multilayers at 60°C and $h_{rel} \approx 95\%$. (*Left column*) intensity integrated along Γ plotted as a function of $q_{\parallel}$. The *vertical dashed line* indicates the position of the sample horizon. (*Right column*) width of the sheet along Γ plotted as a function of $q_{\parallel}$

the best match between simulations and experimental data, which allowed for the calculation of the compression modulus *B* and the bending modulus *κ* of each system (see Sect. 4.1). While *B* is determined by the saccharide-mediated inter-membrane interactions, *κ* depends on the interactions of the molecules in the membrane plane, which are influenced by the saccharide headgroup structure. This is illustrated in Fig. 5.4.

Since refraction and absorption effects, which are not considered in the first Born approximation, become important in the vicinity of the sample horizons (Yoneda wings, see Sect. 2.2.3), these regions were excluded from the comparison of experimental data and simulations. Figure 5.5 shows the integrated intensity (left) and the width (right) of the second Bragg sheet measured from DPPC membrane multilayers at 60 °C and 95% relative humidity, where the multilayers exhibit a lamellar periodicity of $d = 50.9$ Å. As motivated in Sect. 4.1, these representations enable a meaningful comparison between experimental data and simulations based on the free parameters η, λ, and R. In Fig. 5.6, the same is shown

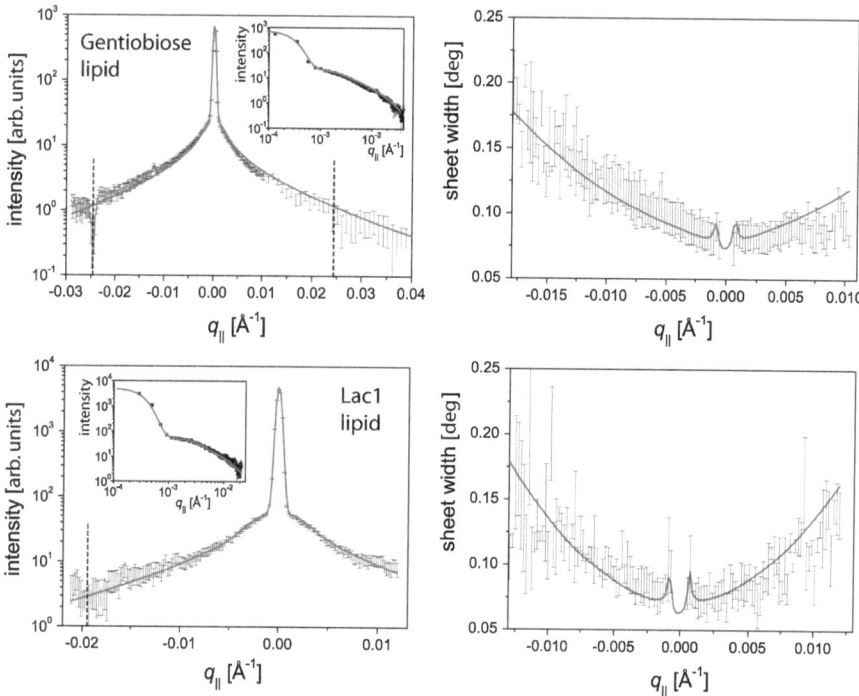

Fig. 5.6 Measured (*data points*) and simulated (*solid line*) second Bragg sheets of gentiobiose lipid (*top*) and Lac1 lipid (*bottom*) membrane multilayers at 80°C and $h_{rel} \approx 95\%$. (*Left column*) intensity integrated along Γ plotted as a function of $q_{\parallel}$ (inset log/log plot). *Vertical dashed lines* indicate the position of the sample horizons. (*Right column*) width of the sheet along Γ plotted as a function of $q_{\parallel}$

for gentiobiose lipid (top) and Lac1 lipid (bottom) membrane multilayers at 80 °C and 95% relative humidity. At these conditions the gentiobiose lipid and Lac1 lipid multilayers assume lamellar periodicities of $d = 50.1$ Å and $d = 56.5$ Å, respectively. The difference in membrane periodicity seems to reflect the difference in the projected length of "bent" and "cylindrical" head groups normal to the membrane plane. It should be noted that data could be recorded not only in reflection but also in transmission (see Sect. 3.3.2), due to the weak neutron absorption of silicon, which enhances the range in which experiments and models can be compared.

DPPC and glycolipid multilayers exhibit a very high alignment with the planar substrate, as can be seen in Figs. 5.5 and 5.6 (left panels) from the sharpness of the central specular maximum (angular width $\approx 0.1°$). The modeled signals (solid red lines) corresponding to the best matching parameters are superimposed on the experimental data points. The best matching model parameters are summarized in Tables 5.1 and 5.2. For all systems the simulations based on the corresponding parameter sets agree well with experimental results in both representations (see Figs. 5.5, 5.6). To highlight the excellent agreement over several orders of

Table 5.1 Parameters of the best matching model for DPPC membrane multilayers at $T = 60$ °C and $h_{rel} \approx 95\%$

System	σ (Å)	η	λ (Å)	R (μm)	κ (J)	κ ($k_B T$)	B (MPa)
DPPC-D	2.2	0.010	6	0.5	9×10^{-20}	18	46

Table 5.2 Parameters of the best matching models for gentiobiose lipid and Lac1 lipid membrane multilayers at $T = 80$ °C and $h_{rel} \approx 95\%$

System	σ (Å)	η	λ (Å)	R (μm)	κ (J)	κ ($k_B T$)	B (MPa)
Gent-D	3.5	0.023	6	0.8	4×10^{-20}	8	21
Lac1-D	2.6	0.011	13	0.5	16×10^{-20}	32	17

magnitude in intensity and $q_{\parallel}$, Fig. 5.6 additionally shows the double-logarithmic presentations of the integrated Bragg sheet intensities as insets.

The obtained R values are in the range of 1 μm, in reasonable agreement with the actual size of the multilayer domains determined by AFM measurements (see Sect. 3.2.2). The obtained bending modulus of DPPC membranes ($\kappa \approx 18\ k_B T$) is in excellent agreement with literature on the mechanical properties of commonly studied phospholipids in fluid L_α-phase [4, 5]. The values obtained for gentiobiose lipid ($\kappa \approx 8\ k_B T$) and Lac1 lipid ($\kappa \approx 32\ k_B T$) have the same order of magnitude, but show a significant difference as a result of the difference in the saccharide headgroup structure. The much higher bending rigidity of Lac1 indicates that the "cylindrical" lactose head groups resist the bending of the membranes much more than "bent" gentiobiose head groups. This seems to coincide with previous studies, which reported that Lac1 lipids form physical gels of ordered head groups [2, 6, 7]. The obtained inter-membrane compression modulus of gentiobiose lipid membranes, $B \approx 21$ MPa, is slightly higher than that of Lac1 lipid membranes ($B \approx 17$ MPa), indicating that the inter-membrane confinement of Lac1 lipid membranes is softer than that of gentiobiose lipid membranes. More strikingly, both values are significantly lower than that obtained for interacting DPPC at the same humidity ($B \approx 46$ MPa), which demonstrates that the disaccharide headgroups lead to much softer inter-membrane interactions than zwitterionic phosphatidylcholine headgroups.

5.1.3.1 Glycolipid Multilayers under Bulk Water

To study the inter-membrane interaction and bending rigidity of the glycolipid membranes in the absence of external osmotic pressures, experiments on gentiobiose lipid membranes in bulk D_2O at 60 °C were carried out using a self developed-liquid cell (see Sect. 3.3.3.2). The thermal stability of the sample during the measurement in bulk water was verified from the symmetry of the Bragg sheets in Ω-direction, since the Ω-axis is proportional to the time axis in a rocking scan. From the Bragg peak positions the periodicity of the gentiobiose lipid membrane multilayers was determined to be $d = 56.8$ Å, which is in good agreement with the

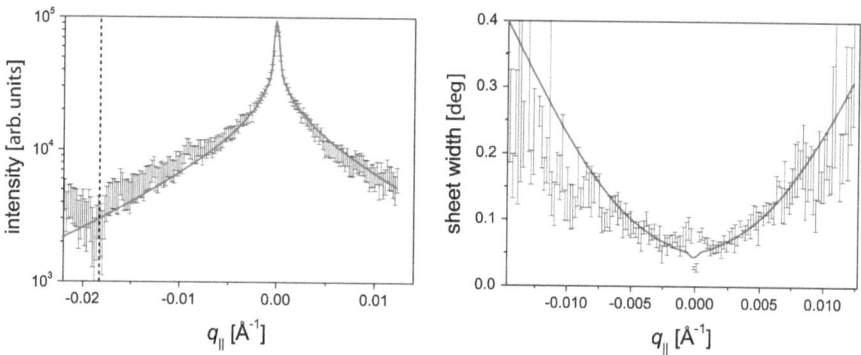

Fig. 5.7 Measured (data points) and simulated (*solid line*) second Bragg sheets of gentiobiose lipid membrane multilayers at 60°C in bulk water. (*Left column*) intensity integrated along Γ plotted as a function of $q_{\parallel}$. The *vertical dashed line* indicates the position of the sample horizon. (*Right column*) width of the sheet along Γ plotted as a function of $q_{\parallel}$

Table 5.3 Parameters of the best matching model for gentiobiose lipid membrane multilayers at $T = 60$ °C under bulk water

System	σ (Å)	η	λ (Å)	R (μm)	κ (J)	κ ($k_B T$)	B (MPa)
Gent-H	8.4	0.10	24	1.4	3.2×10^{-20}	7	0.9

value obtained by SAXS at 70 °C, $d = 58$ Å [8]. This value is rather small compared to the bulk water spacing of DPPC multilayers, which is around 65 Å (see Sect. 5.2.1). The difference can be understood from the short characteristic hydration decay length found for gentiobiose lipid (see Sect. 5.1.2). Figure 5.7 shows the measured and simulated second Bragg sheet of gentiobiose lipid in bulk water. The parameters corresponding to the best matching model (red solid line in Fig. 5.7) are presented in Table 5.3.

Interestingly, the obtained bending modulus ($\kappa_{bulk} \approx 7\, k_B T$) is similar to that obtained at high humidity ($\kappa_{vapor} \approx 8\, k_B T$), indicating that the work required for bending the membranes is not influenced significantly by the presence of bulk water between the neighboring membranes. On the other hand, the inter-membrane compression modulus, $B_{bulk} \approx 0.9$ MPa, is 20-fold lower than the corresponding value under osmotic stress, $B_{vapor} \approx 21$ MPa and seems to reflect the softer inter-membrane confinement in the presence of a water interlayer. Both tendencies are in good agreement with previous studies on neutral and charged phospholipid membranes [9, 10].

5.1.4 Summary of Sect. 5.1

It was demonstrated that solid-supported glycolipid membrane multilayers constitute a well-defined model system for the study of the mechanical properties of interacting glycolipid membranes by specular and off-specular neutron scattering. The influence of a distinct difference in the saccharide structure on inter-membrane interactions and membrane mechanics was investigated. Force–distance relationships recorded in scattering experiments under various defined osmotic pressures revealed a clear difference in the steric compressibilities of the membrane-bound gentiobiose and lactose layers at high osmotic pressures. At lower osmotic pressures membrane multilayers of both glycolipids exhibited a much weaker swelling than typical for commonly studied phospholipid systems. This finding suggests that attraction between saccharide headgroups contributes to the inter-membrane interactions. The bending rigidities of gentiobiose lipid and Lac1 lipid membranes, determined from the off-specular scattering signals, were found to be significantly different, which indicates that in contrast to "bent" gentiobiose headgroups, "cylindrical" lactose headgroups have a considerable contribution to the overall bending rigidity of the membranes. Measurements of gentiobiose lipid membrane multilayers in bulk water showed that the water layer between the membranes has no significant influence of the bending rigidity but strongly softens the inter-membrane confinement.

5.2 Role of Specific Saccharide–Saccharide Interactions in Membrane–Membrane Contacts

The homophilic interaction of LewisX saccharide motifs is involved in the early embryonic development [11] and has been studied extensively using a variety of techniques, such as NMR spectroscopy [12–15] and atomic force microscopy [16] (AFM). In this thesis, glycolipids bearing the LewisX trisaccharide (Le^X lipid, see Sect. 3.1.1) were incorporated at defined molar fractions into solid-supported multilayers (see Sect. 2.1.4.4) of phospholipid membranes, in order to study the specific homophilic interactions of the LewisX motifs. DPPC was chosen as matrix lipid, as it possesses the same type of all-saturated hexadecyl chains as the Le^X lipid. All measurements were carried out at $T = 60$ °C, a temperature where matrix lipids are in fluid L_α-phase and thus the lateral mobility of the Le^X lipid is guaranteed. The model system is schematically illustrated in Fig. 5.8 (left) and was studied using specular and off-specular neutron scattering under various buffer conditions. The right panel of the figure shows a reciprocal space map recorded with DPPC membrane multilayers doped with 25 mol% Le^X lipid in *Ca-free NaCl buffer*. To gain maximum contrast in scattering length density, chain deuterated Le^X lipid was used in combination with chain-deuterated DPPC (see Sect. 3.1), and all buffers were based on pure H_2O. Although during the measurements the samples were immersed in bulk buffer for several hours at high temperature (60 °C), they showed excellent stability (verified from the symmetry of the Bragg sheets, see Sect. 5.1.3) and the scattering signals were remarkably strong (with excellent statistics and signal/noise ratio, see Fig. 5.8, right).

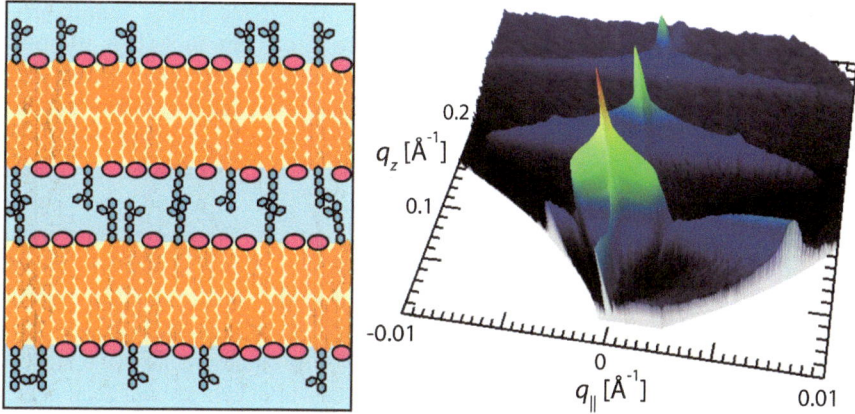

Fig. 5.8 : (*left*) Sketch of the studied oriented DPPC membrane multilayers doped with Le^X lipid. Phosphatidylcholine headgroups are indicated with ellipses and saccharide units with hexagons. (*right*) Reciprocal space map recorded by specular and off-specular neutron scattering from DPPC doped with 25 mol% Le^X lipid at 60 °C in *Ca-free NaCl buffer*

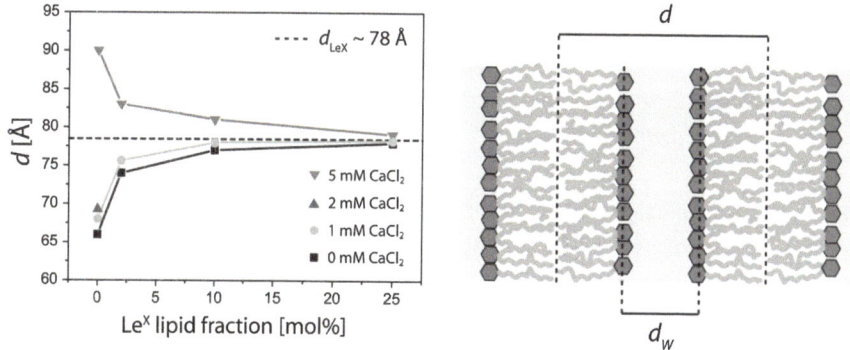

Fig. 5.9 Lamellar periodicities d of DPPC membrane multilayers doped with various molar fractions of LeX lipid at various calcium concentrations (*left*). Sketch of interacting phospholipid membranes with the definitions of d and d$_W$ (*right*)

5.2.1 Influence of LewisX Trisaccharides on Inter-Membrane Interactions

Solid-supported membranes multilayers of DPPC, either pure or doped with defined molar fractions of Le^X lipid were investigated in buffers with various concentrations of CaCl$_2$. The equilibrium lamellar periodicities d were calculated from the q_z-positions of the Bragg peaks in the scattering signals. The obtained values are summarized in Fig. 5.9 (left). It is seen that calcium ions have a substantial influence on the periodicity of pure DPPC membrane multilayers. The values range from $d = 66.1$ Å in the absence of calcium to 90.0 Å at 5 mM $CaCl_2$. This well-understandable effect has its origin in electrostatics and is discussed in detail in Sect. 5.2.2.2. Interestingly, at all studied calcium concentrations (0, 1, and 5 mM), the incorporation of Le^X lipid significantly alters the multilayer periodicity. Most prominently, the values converge with increasing Le^X lipid concentration to a saturation periodicity of about $d_{LeX} \approx 78$ Å (indicated with a dashed line in Fig. 5.9, left), which seems to correspond to the favored membrane separation for the specific trans-homophilic LewisX–LewisX interaction. In fact, this periodicity appears very reasonable considering the thickness of the hydrocarbon chain region (≈ 40 Å) plus the length of two saccharide headgroups (≈ 22 Å each) that slightly interdigitate when they form a pair. While at the highest studied Le^X lipid concentration (25 mol%) the influence of calcium on the lamellar periodicity almost disappears (≈ 1 Å), at the lowest studied finite Le^X lipid concentration (2 mol%) the spacing deviates from d_{LeX} by up to 5 Å to both directions depending on the calcium concentration. This can be well understood from the conformational flexibility of the Le^X lipid molecules with pentasaccharide headgroups. Nonetheless, the multilayer periodicity is substantially influenced even by this low Le^X lipid fraction, suggesting that the membranes are cross-linked by the trans-homophilic saccharide–saccharide interactions.

5.2.2 Specific Saccharide–Saccharide Interactions under Compressional or Tensile Stress

At low calcium concentrations (0 mM, 1 mM), where the periodicities favored by the DPPC matrix ($d = 66.1$ Å and $d = 67.9$ Å, respectively) are smaller than d_{LeX}, LewisX pairs are subject to compressional forces. Conversely, LewisX pairs are subject to tensile forces at 5 mM calcium, where pure DPPC membranes exhibit a spacing of $d = 90.0$ Å $> d_{LeX}$. Regarding this fact, the use of calcium offers the unique possibility to study the specific LewisX–LewisX interaction under defined forces. For a quantitative discussion of the compressional and tensile forces exerted to the LewisX pairs under given buffer conditions, the interaction potential of the membranes is modeled in the following.

5.2.2.1 Interactions of the Matrix (DPPC) Membranes

In the first step, the inter-membrane interactions of the matrix lipid (DPPC) membranes are quantified. This is a prerequisite for the detailed investigation of the specific inter-membrane interactions of the LewisX saccharide motifs. As motivated in Sect. 5.1.2, the comprehensive theoretical description of all relevant inter-membrane interactions is very challenging in general. However, DPPC membranes are commonly studied model systems, which have been characterized with high detail [17, 18]. From this starting point, a quantitative description of the inter-membrane interactions is feasible (see Sect. 2.1.2). The interaction of charge-neutral DPPC membranes in L_α-phase in bulk water (i.e., in the absence of calcium ions) can be modeled in good approximation with a suited definition of the membrane separation d_W (see Fig. 5.9) and considering three force contributions:

1. *Van de Waals (vdW) attraction* According to Sect. 2.1.2.2, d_W is defined as the lamellar periodicity d minus the "hydrophobic membrane thickness" d_H. This approach has been previously used successfully for DMPC [19]. In this thesis, $d_H = 38.7$ Å was used, calculated from the value reported by Janiak [20, 21] for DMPC (with tetradecyl chains) in L_α-phase ($d_H = 35.5$ Å) by correcting for the longer hydrocarbon chains of DPPC (with hexadecyl chains). The correction corresponds to four times the average projected length of a hydrocarbon C–C bond in L_α-phase membranes [4] (0.8 Å).
2. *Hydration repulsion* d_W is used in the same definition as for the *vdW* attraction, together with the parameters $\lambda_{hyd} = 2.0$ Å, and $\Pi_0 = 4.5 \times 10^9$ Pa (see Sect. 2.1.2.3).
3. *Undulation repulsion* d_W was defined as d minus the "steric thickness" d_S of DPPC membranes in L_α-phase. $d_S = 45.9$ Å was used, as reported by Kucerka and Nagle [17]. In the calculations the prefactor $\alpha = 0.104$ (see Sect. 2.1.2.4) was taken from Bachmann [22], together with the above presented result for the bending rigidity of DPPC at 60°C in L_α-phase, $\kappa = 18$ k_BT (see Sect. 5.1.3).

In Fig. 5.10 (left), the modeled interaction contributions are plotted as a function of the lamellar periodicity d. As can be seen, for low d-values, corresponding to short inter-membrane separations, the interaction is dominated by hydration repulsion, while for larger separation the *vdW* attraction is dominant. The right panel of the figure shows the absolute of the modeled disjoining pressure Π (i.e., the sum of all three force contributions, see Sect. 2.1.2.6). Π is zero at $d = 66.0$ Å (see Fig. 5.10, right). This spacing corresponds to the equilibrium separation of DPPC membranes predicted by the model for the absence of calcium ions. Lower separations lead to repulsion, while larger separations lead to attraction. In other words, the membranes are confined at $d = 66.0$ Å in a stable potential minimum with $\frac{\partial \Pi}{\partial d_W} < 0$ (see Sect. 2.1.2.6). In *Ca-free NaCl buffer* the experimentally obtained lamellar periodicity of the DPPC membranes is $d = 66.1$ Å (see also Fig. 5.9), which is in excellent agreement with the value of $d = 66.0$ Å theoretically derived above.[1]

---

[1] In pure H_2O, $d \approx 65$ Å is observed. The influence of monovalent salt ($\Delta d \approx 1$ Å for 100 mM NaCl) follows the trend predicted by Petrache et al. [27], but is very weak. Therefore, this effect is neglected in the following considerations.

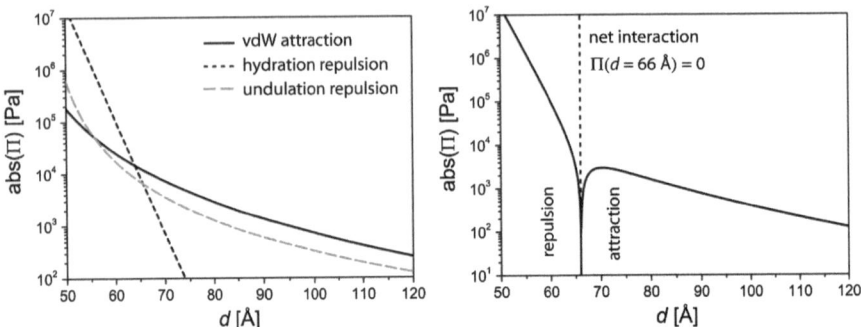

Fig. 5.10 Modeled interactions of pure DPPC membranes in water. individual force contributions (*vdW* attraction, hydration repulsion, and undulation repulsion) (*left*). Absolute of the net interaction Π which vanishes at the predicted equilibrium lamellar periodicity (*right*)

5.2.2.2 Modification of Matrix Membrane Interactions with Calcium Ions

Several studies have reported that calcium ions have a strong influence on the equilibrium separation of phospholipid membranes. It was found in diffraction experiments that the lamellar periodicity of the membranes increases dramatically upon the addition of calcium [23, 24]. Deuterium magnetic resonance experiments have evidenced the adsorption of calcium to phosphatidylcholine headgroups [25]. Based on these results, a scenario where the membrane surfaces become positively charged due to the adsorption of divalent cations resulting in an electrostatic repulsion has become consensus view. As shown above (Fig. 5.9), a major influence of calcium on DPPC multilayers is observed in this thesis: Already the addition of 1 mM $CaCl_2$ causes an increase in lamellar spacing by $\Delta d = 1.8$ Å ($d = 67.9$ Å). At 5 mM $CaCl_2$ the spacing is increased by as much as $\Delta d = 23.9$ Å ($d = 90.0$ Å). An overview is given in Table 5.4. In the following, the effect of calcium is quantified with Δd by subtracting the reference value ($d = 66.1$ Å) obtained in *Ca-free NaCl buffer*.

To account theoretically for the effect of calcium, electrostatic repulsion has to be included into the force balance that determines the predicted equilibrium periodicity. For this purpose the calcium ions adsorbed to the membranes were represented as homogeneously charged plates located at the center of the PC headgroup. Accordingly, d_W was defined as d minus the headgroup–headgroup distance d_{HH} of DPPC for the description of the electrostatic interaction. Here,

Table 5.4 Lamellar periodicities of DPPC multilayers subject to various calcium concentrations

Medium	Calcium concentration (mM)	d (Å)	Δd (Å)
Ca-free NaCl buffer	0	66.1	0.0
Ca01 NaCl buffer	1	67.9	1.8
Ca02 NaCl buffer	2	69.1	3.0
Ca05 NaCl buffer	5	90.0	23.9

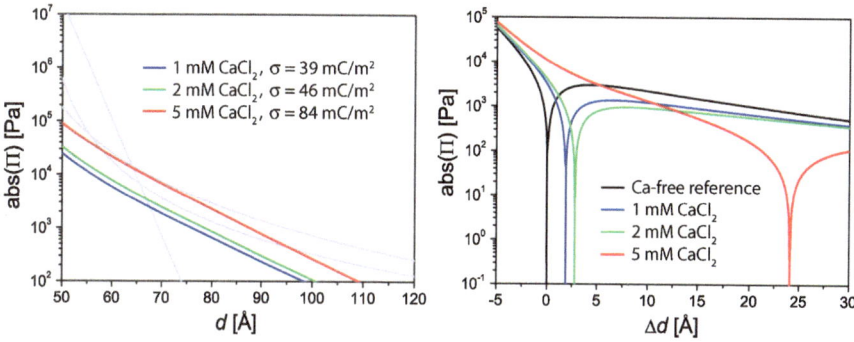

Fig. 5.11 Modeled electrostatic interactions in calcium-loaded buffers. The other force contributions are shown in the same plot in light grey (*left*). The points of vanishing disjoining pressure represent the resulting change Δd in equilibrium lamellar periodicity (*right*)

Table 5.5 Characteristics of DPPC membranes subject to various calcium concentrations: charge densities, number of DPPC molecules per adsorbed calcium ion, and calcium-DPPC binding affinity constants

Calcium concentration (mM)	Charge density, σ (mC/m^2)	# DPPC per calcium ion	Affinity, K (M^{-1})
1	39	126	7.9
2	46	107	4.6
5	84	59	3.3

$d_{HH} = 37.8$ Å was used, according to the results of Kucerka and Nagle [17]. As discussed in Sect. 2.1.2, the correct description of electrostatic interactions in mixed electrolytes (here NaCl and CaCl$_2$) is non-trivial. The numerical calculations involved are discussed in Sect. 4.2. The positive membrane surface charge density σ was used as a free parameter which was adjusted such that the predicted shifts in lamellar spacing matched the experimentally observed values of Δd. Figure 5.11 (left) shows the strength of the modeled electrostatic repulsion according to the σ-values obtained to match Δd for 1, 2, and 5 mM CaCl$_2$. The right panel of the figure shows the absolute of the disjoining pressures as they result from the summation of all modeled force contributions. According to the choice of σ, the curves exhibit zero-crossings at Δd values matching the experimental results.

A clear trend of increasing charge density with increasing calcium concentration is observed (from $\sigma \approx 39$ mC/m^2 at 1 mM CaCl$_2$ to ≈ 84 mC/m^2 at 5 mM CaCl$_2$). This is summarized in Table 5.5. For each calcium concentration, the corresponding numbers of DPPC molecules per adsorbed calcium ion were calculated from the charge densities with the assumption that one DPPC molecule covers a membrane area of 65 Å^2 in fluid L_α-phase [26]. Following the recipe of Petrache et al. [27]., the binding constant K of the DPPC / Ca^{2+} interaction was estimated:

$$K = [DPPC \oplus Ca^{2+}]/([DPPC][Ca^{2+}]),$$

where $[DPPC \oplus Ca^{2+}]$ denotes the molar fraction of DPPC molecules with a bound calcium ion, $[DPPC]$ the molar fraction of un-complexed DPPC molecules, and $[Ca^{2+}]$ the calcium concentration in M. The calculated values are summarized in Table 5.5. The binding affinities are seen to be quite weak, at the order of several M^{-1}. Furthermore, a clear trend of decreasing affinity with increasing calcium concentration is observed. This decrease can be understood considering the higher charge density of the membrane surface at higher calcium concentrations, which would render the adsorption of Ca^{2+} less favorable.

5.2.2.3 Cross-Linking the Membranes: Forces and Energies

To estimate the forces and energies involved in the coupling of neighboring membranes by specific carbohydrate–carbohydrate interactions, DPPC membranes doped with 2 mol% Le^X lipid at 5 mM $CaCl_2$ are discussed in detail. Here, the presence of 2 mol% Le^X lipid causes a reduction in multilayer periodicity from $d_{0\%}$ (5 mM) $= 90.0$ Å to $d_{2\%}$ (5 mM) $= 83.0$ Å $> d_{LeX} \approx 78$ Å. The LewisX pairs are subject to tensile forces, and thus any effects of steric repulsion, originating from the LewisX headgroups, can be excluded. Due to the low molar fraction of Le^X lipid, the non-specific inter-membrane interactions are dominated by the matrix lipids in this system and can be approximated by the disjoining pressure profile modeled for pure DPPC membranes at 5 mM $CaCl_2$, as shown in Fig. 5.11. This curve is evaluated at $d = d_{2\%}$(5 mM) $= 83.0$ Å, in order to quantify the repulsive pressure $\Pi_{2\%}$ (5 mM) that has to be compensated by the specific LewisX–LewisX coupling at 5 mM $CaCl_2$. $\Pi_{2\%}$ (5 mM) is found to be 351 Pa (see Fig. 5.12). Along the same line, the mechanical work per unit area, $\omega_{2\%}$ (5 mM) that has to be performed by the specific LewisX–LewisX interactions

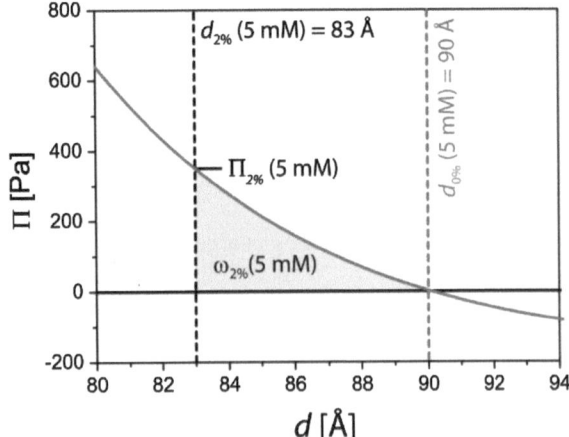

Fig. 5.12 Modeled pressure $\Pi_{2\%}$(5 mM) and mechanical work $\omega_{2\%}$(5 mM) to be overcome by the specific LewisX interactions at 2 mol% Le^X lipid and 5 mM $CaCl_2$ in order to decrease the lamellar periodicity by the experimentally observed increment [from $d_{0\%}$(5 mM) = 90.0 Å to $d_{2\%}$(5 mM) = 83.0 Å]. The *solid curve* represents the disjoining pressure profile modeled for pure DPPC membranes at 5 mM $CaCl_2$

in order to decrease the lamellar spacing from $d = d_{0\%}(5 \text{ mM}) = 90.0 \text{ Å}$ to $d = d_{2\%}(5 \text{ mM}) = 83.0 \text{ Å}$ can be estimated by integrating the curve in Fig. 5.12 between these spacings (see shaded area), which yields $\omega_{2\%}$ (5 mM) = 110 nJ/m^2. By assuming that all available LewisX motifs are involved in the membrane cross-linking, a lower estimate of the force and energy per pair can be given as $F \approx 10$ fN and $E \approx 0.001$ $k_B T$, respectively. However, it is likely that in average only a sub-fraction of LewisX motifs is coupled in pairs, and thus typical forces and energies per pair are higher.

The binding energy of LewisX pairs was reported [16, 28] to be at the order of several $k_B T$. Rupture forces at the order of 10 pN were reported by Tromas et al. [16]. Considering the low energies and forces obtained in this thesis, very low densities of LewisX pairs (corresponding to one pair per 1,000 s of square nanometers) are expected to be sufficient for the cross-linking of phospholipid membranes under physiological conditions. There are several studies [12–15] suggesting that the LewisX–LewisX interaction is mediated by divalent cations—preferably Ca^{2+}. In contrast to these findings, there is evidence that homophilic LewisX interaction occurs also without complexing Ca^{2+}. For example, in an AFM experiment the presence of Ca^{2+} did not contribute significantly to the binding forces [15]. It was speculated that in nature Ca^{2+} may only be responsible for the approach and organization of the carbohydrates in the cell membrane. In further experiments, LewisX aggregates were detected in the absence of Ca^{2+} by cold spray ionization [29], and molecular dynamics simulations [28] suggested that the LewisX dimer is stable with or without Ca^{2+}, but that the dimers are strengthened by Ca^{2+}. In this thesis, it was seen that Ca^{2+} is not required for the formation of *trans*-homophilic LewisX pairs. However, the here presented experimental results cannot exclude a strengthening of the specific interaction, as the tensile forces exerted to the pairs in this study are likely too low to resolve the possible difference.

5.2.3 Influence of LewisX on the Mechanics of Membrane Multilayers

To reveal the influence of *trans*-homophilic LewisX pairs on the mechanical properties of interacting DPPC membranes, the off-specular scattering signals were analyzed (see Sect. 4.1). Here, a comparison between DPPC multilayers doped with 10 and 25 mol% Le^X lipid in *Ca-free NaCl buffer* is shown. Both systems exhibit lamellar spacings very close to the "saturation value" of $d_{LeX} \approx 78$ Å. Thus, differences in the obtained mechanical parameters can be directly related to the different lateral densities of Le^X lipid molecules, while avoiding influences borne from a difference in the membrane separation. Figure 5.13 shows the integrated intensities (left) and the widths (right) of the second Bragg sheets measured from DPPC multilayers doped with 10 mol% (top) and 25 mol% (bottom) Le^X lipid. The modeled signals (solid red lines)

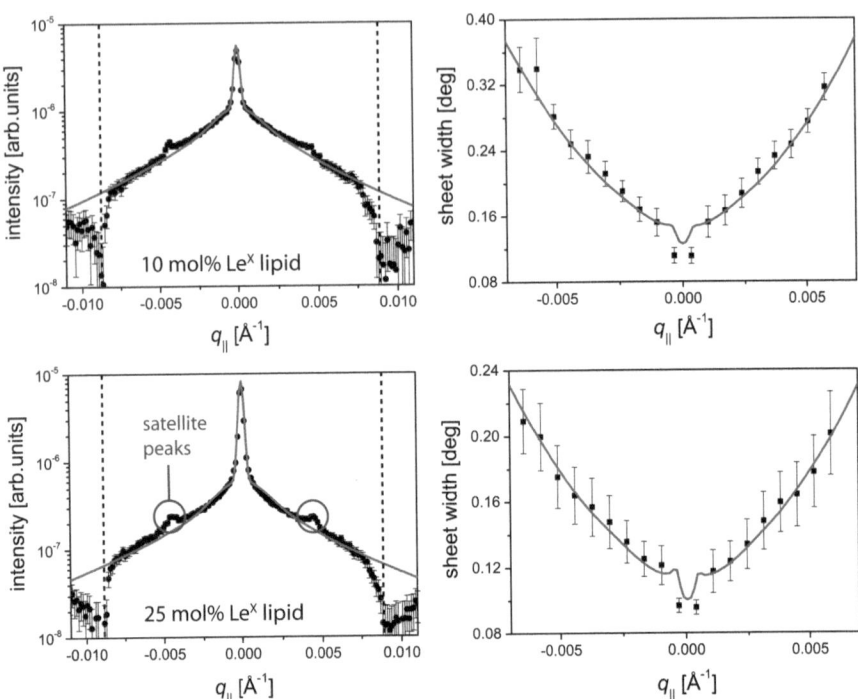

Fig. 5.13 Measured (*data points*) and simulated (*solid line*) second Bragg sheets of DPPC membrane multilayers doped with 10 mol% (*top*) and 25 mol% (*bottom*) Le^X lipid at 60°C in Ca-free NaCl buffer. (*Left column*) intensity integrated along Γ plotted as a function of $q_\parallel$. Vertical *dashed lines* indicate the position of the sample horizons, separating the reflection regime (*center*) from the transmission regimes. (*Right column*) width of the sheet along Γ plotted as a function of $q_\parallel$

corresponding to the best matching parameters are superimposed on the experimental data points. Here, the rather thick (0.1 mm) layer of H_2O-based buffers in the liquid cell leads to strong incoherent scattering by ^1H and results in a high absorption of the neutron beam in the transmission regime. Therefore, only the reflection regime (between the vertical dashed lines in Fig. 5.13, left panels) was used for the comparison between experimental and modeled signals. Another feature of the scattering signals not captured by the models is the existence of satellite peaks (indicated in the bottom left panel). From their position in q_z and $q_\parallel$, they can be clearly identified as results of double scattering processes located at

$$q_z \cong \frac{4\pi}{d} \quad \text{and} \quad q_\parallel \cong \pm\frac{2\pi}{\lambda}\left(\frac{\lambda}{d}\right)^2 .$$

Therefore experimental and modeled signals are not compared in the close vicinity of the satellite peaks. Despite these well-understood and predictable discrepancies, the models are in excellent agreement with the experimental data in the valid range. Most strikingly, the almost perfect parabolic shapes of the sheet

Table 5.6 Parameters of the best matching models for DPPC membrane multilayers doped with 10 mol% and 25 mol% Le^X lipid at 60 °C in Ca-free NaCl buffer

Le^X lipid fraction (mol%)	σ (Å)	η	λ (Å)	R (μm)	κ (J)	κ (k_BT)	B (MPa)
10	9.6	0.085	65	0.8	8×10^{-20}	16	0.22
25	7.1	0.046	38	0.7	8×10^{-20}	17	0.70

widths (right panels) enable the precise experimental determination of the de Gennes parameter λ, which significantly improves the accuracy with which the mechanical parameters B and κ can be measured.

The best-matching model parameters are summarized in Table 5.6. The obtained cut-off radii R are in the range of 1 μm, consistent with those found for other studied multilayer systems (see Sect. 5.1.3). Interestingly, the bending rigidities obtained with both Le^X lipid fractions ($\kappa \approx 16\ k_BT$ for 10 mol% and $\kappa \approx 17\ k_BT$ for 25 mol%) are almost identical to the one determined for pure DPPC membranes ($\kappa \approx 18\ k_BT$, see Sect. 5.1.3). This indicates that the lateral density of membrane-anchored pentasaccharides is not high enough to contribute significantly to the membrane bending rigidity via in-plane steric interactions, even at the highest studied Le^X lipid density (25 mol%). On the other hand, the compression moduli obtained for 10 mol% ($B = 0.22$ MPa) and 25 mol% ($B = 0.70$ MPa) Le^X lipid show a clear difference: the higher density of LewisX results in a much stronger confinement of the membranes between their neighbors. This behavior supports the interpretation that the membranes are cross-linked with *trans*-homophilic LewisX pairs, where an increase in the number of cross-links is expected to result in a stronger confinement. This result is analogous to the above discussed influence of calcium on the lamellar spacing of DPPC membrane multilayers doped with Le^X lipid: For higher LewisX densities, the lamellar spacing is kept closer to the optimum value ($d_{LeX} = 78$ Å), despite the compressional or tensile forces exerted by the DPPC matrix under various calcium concentrations.

5.2.4 Summary of Sect. 5.2

It was demonstrated that solid-supported membrane multilayers doped with membrane-anchored carbohydrates serve as a unique platform for the study of specific carbohydrate–carbohydrate interactions under biological conditions. Calcium ions were utilized to exert defined compressional or tensile forces to the systems, while specular and off-specular neutron scattering provided the simultaneous determination of structure and mechanics. It was shown that LewisX forms specific homophilic pairs that cross-link adjacent membranes. The lamellar periodicity of the multilayers was found to approach to the value preferred by the specific pairs with increasing Le^X lipid density. A theoretical estimation of forces and energies required to cross-link the neighboring membranes was conducted and indicated that the membranes are cross-linked even at the lowest studied Le^X lipid density, consistently with the experimental results. The mechanical parameters extracted from the off-specular scattering signals showed that LewisX has no significant influence on the bending rigidity of the matrix membrane throughout the studied Le^X lipid densities. In contrast, the inter-membrane confinement was found to increase with increasing density of cross-linking Le^X lipid.

References

1. M.F. Schneider, R. Zantl, C. Gege, R.R. Schmidt, M. Rappolt, M. Tanaka, Hydrophilic/hydrophobic balance determines morphology of glycolipids with oligolactose headgroups. Biophys. J. **84**, 306 (2003)
2. M. Tanaka, S. Schiefer, C. Gege, R.R. Schmidt, G.G. Fuller, Influence of subphase conditions on interfacial viscoelastic properties of synthetic lipids with gentiobiose head groups. J. Phys. Chem. B **108**, 3211 (2004)
3. G. Cevc, *Phospholipids handbook* (Marcel Dekker, New York, 1993)
4. E. Sackmann, in *Structure and Dynamics of Membranes*, ed. by R. Lipowski, E. Sackmann (Elsevier, Amsterdam, 1995)
5. J. Daillant, E. Bellet-Amalric, A. Braslau, T. Charitat, G. Fragneto, F. Graner, S. Mora, F. Rieutord, B. Stidder, Structure and fluctuations of a single floating lipid bilayer. Proc. Natl. Acad. Sci. USA **102**, 11639 (2005)
6. M. Tanaka, M.F. Schneider, G. Brezesinski, In-plane structures of synthetic oligolactose lipid monolayers—impact of saccharide chain length. Chem. Phys. Chem. **4**, 1316 (2003)
7. M.F. Schneider, K. Lim, G.G. Fuller, M. Tanaka, Rheology of glycocalix model at air/water interface. Phys. Chem. Chem. Phys. **4**, 1949 (2002)
8. M. Tanaka, F. Rehfeldt, S.S. Funari, C. Gege, R. R. Schmidt, HASYLAB Annual Report, 2004
9. G. Brotons, L. Belloni, T. Zemb, T. Salditt, Elasticity of fluctuating charged membranes probed by X-ray grazing-incidence diffuse scattering. Europhys. Lett. **75**, 992 (2006)
10. J. Pan, S. Tristram-Nagle, N. Kucerka, J.F. Nagle, Temperature dependence of structure, bending rigidity, and bilayer interactions of dioleoylphosphatidylcholine bilayers. Biophys. J. **94**, 117 (2008)
11. I. Eggens, B. Fenderson, T. Toyokuni, B. Dean, M. Stroud, S. Hakomori, Specific interaction between Lex and Lex determinants. A possible basis for cell recognition in preimplantation embryos and in embryonal carcinoma cells. J. Biol. Chem **264**, 9476 (1989)
12. C. Gege, A. Geyer, R.R. Schmidt, Synthesis and molecular tumbling properties of sialyl Lewis X and derived neoglycolipids. Chem. Eur. J. **8**, 2454 (2002)
13. A. Geyer, C. Gege, R.R. Schmidt, Carbohydrate–carbohydrate recognition between LewisX glycoconjugates. Angew. Chem. Int. Ed. **38**, 1466 (1999)
14. A. Geyer, C. Gege, R.R. Schmidt, Calcium-dependent carbohydrate-carbohydrate recognition between Lewis^X blood group antigens. Angew. Chem. Int. Ed. **39**, 3246 (2000)
15. G. Nodet, L. Poggi, D. Abergel, C. Gourmala, D. Dong, Y. Zhang, J.M. Mallet, G. Bodenhausen, Weak calcium-mediated interaction between LewisX-related trisaccharides studied by NMR measurements of residual dipolar couplings. J. Am. Chem. Soc. **129**, 9080 (2007)
16. C. Tromas, J. Rojo, J. M. de la Fuente, A.G. Barrientos, R. García, S. Penadés, Adhesion forces between Lewis X determinant antigens as measured by atomic force microscopy, Angew. Chem. Int. Ed. **40**, 3052 (2001)
17. N. Kucerka, S. Tristram-Nagle, J.F. Nagle, Closer look at structure of fully hydrated fluid phase DPPC bilayers. Biophys. J. **90**, L83 (2006)

18. J.F. Nagle, R. Zhang, S. Tristram-Nagle, W. Sun, H.I. Petrache, R.M. Suter, X-ray structure determination of fully hydrated L-alpha phase dipalmitoylphosphatidylcholine bilayers. Biophys. J. **70**, 1419 (1996)
19. B. Demé, M. Dubois, T. Zemb, Swelling of a lecithin lamellar phase induced by small carbohydrate solutes. Biophys. J. **82**, 215 (2002)
20. M.J. Janiak, M.D. Small, G.G. Shipley, Nature of the thermal pretransition of synthetic phospholipids: dimyristoyl- and dipalmitoyl lecithin. Biochemistry **15**, 4575 (1976)
21. M.J. Janiak, M.D. Small, G.G. Shipley, Temperature and compositional dependence of the structure of hydrated dimyristoyl lecithin. J. Biol. Chem. **254**, 6068 (1979)
22. M. Bachmann, H. Kleinert, A. Pelster, Fluctuation pressure of a stack of membranes. Phys. Rev. E **63**, 051709 (2001)
23. L.J. Lis, W.T. Lis, V.A. Parsegian, R.P. Rand, Adsorption of divalent cations to a variety of phosphatidylcholine bilayers. Biochemistry **20**, 1771 (1981)
24. L.J. Lis, V.A. Parsegian, R.P. Rand, Binding of divalent cations to dipalmitoylphosphatidyl-choline bilayers and its effect on bilayer interaction. Biochemistry **20**, 1761 (1981)
25. C. Altenbach, J. Seelig, Calcium binding to phosphatidylcholine bilayers as studied by deuterium magnetic resonance. Evidence for the formation of a calcium complex with two phospholipid molecules. Biochemistry **23**, 3913 (1984)
26. J.F. Nagle, S. Tristram-Nagle, Structure of lipid bilayers. Biochim. Biophys. Acta **1469**, 159 (2000)
27. H.I. Petrache, T. Zemb, L. Belloni, V.A. Parsegian, Salt screening and specific ion adsorption determine neutral-lipid membrane interactions. PNAS **103**, 7982 (2006)
28. C. Gourmala, Y. Luo, F. Barbault, Y. Zhang, S. Ghalem, F. Maurel, B. Fan, Elucidation of the LewisX–LewisX carbohydrate interaction with molecular dynamics simulations: a glycosynapse model. J. Mol. Struct. Theochem **821**, 22 (2007)
29. S.-I. Nishimura, N. Nagahori, K. Takaya, Y. Tachibana, N. Miura, K. Monde, Direct observation of sugar-protein, sugar–sugar, and sugar–water complexes by cold-spray ionization time-of-flight mass spectrometry. Angew. Chem. Int. Ed. **44**, 571 (2005)

Chapter 6
Structure and Mechanical Properties of Bacteria Surfaces

In this chapter, model systems of bacteria surfaces, prepared from lipopolysac-charides (LPSs) of various structural complexities, are investigated using neutron scattering, high-energy X-ray reflectometry, and grazing-incidence X-ray fluo-rescence (GIXF). In particular, the influence of divalent cations on the confor-mation of LPSs and on the mechanics of LPS membranes is studied. As motivated in Chap.1, these effects are considered crucial for the resistance of Gram-negative bacteria against cationic antimicrobial peptides (CAPs), but experimental evidence on the molecular level is still missing.

Since the mechanics of complex LPS membranes and their response to divalent cations can be best understood if the mechanical properties of their basic building blocks are known, interacting membranes of Lipid A and rough mutant LPSs (see Sect. 3.1.2) are investigated in the first part. Their structure and mechanics is determined by specular and off-specular neutron scattering. The influence of saccharide structure, sample hydration, and divalent calcium ions is studied.

In the second part, the impact of divalent ions on the conformation of complex wild-type LPSs with long O-sidechains is investigated by high-energy X-ray reflectivity measurements from a solid-supported LPS monolayer in bulk buffer.

In the third part, GIXF measurements from rough mutant LPS monolayers at the air/water interface are carried out in order to reconstruct the density profiles of ions at LPS surfaces, since the influence of ions on saccharide conformation and membrane mechanics is closely related to their spatial distribution.

6.1 Influence of Lipopolysaccharide Structure and Divalent Cations on the Mechanics of LPS Multilayers

As demonstrated in Chap.5, the study of solid-supported membrane multilayers under controlled osmotic pressures and under bulk buffers by specular and off-

E. Schneck, *Generic and Specific Roles of Saccharides at Cell and Bacteria Surfaces,* Springer Theses, DOI: 10.1007/978-3-642-15450-8_6,
© Springer-Verlag Berlin Heidelberg 2011

 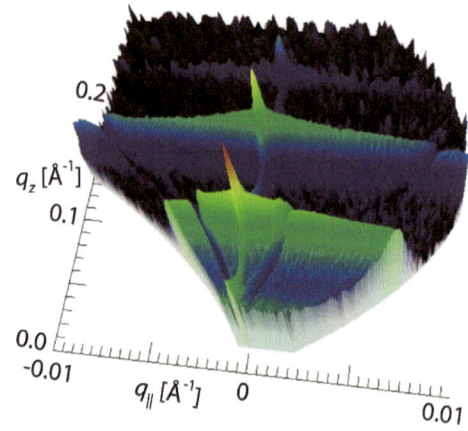

Fig. 6.1 (*Left*) Sketch of oriented rough mutant LPS membrane multilayers. Saccharide units are indicated with hexagons. (*Right*) Reciprocal space map recorded by specular and off-specular neutron scattering from LPS Ra at 60°C and high relative humidity ($\approx 95\%$)

specular neutron scattering enables comprehensive insight into structure and mechanics of the interacting membranes. Thus, the same method was applied to Lipid A and rough mutant LPS membranes. A sketch of interacting LPS membrane in a multilayer stack is shown in Fig. 6.1 (left). The right panel of the figure shows a reciprocal space map recorded with solid-supported LPS Ra membrane multilayers at 60°C and high relative humidity ($\approx 95\%$). The scattering signals shown in this thesis represent the first specular and off-specular neutron scattering experiments with complex and realistic bacteria membrane models.

To ensure that all systems are in fluid L_α-phase throughout the experiments, all measurements were conducted at 60°C, which is more than 10°C above the chain melting temperature of all studied molecules (see Table 3.1). To investigate the influence of molecular chemistry (length and charge of oligosaccharide head groups) on the vertical/lateral structural ordering and mechanical properties of Lipid A and mutant LPS membranes, the scattering experiments were carried out under the following four conditions: (1) at low ($\approx 20\%$) relative humidity (corresponding to an osmotic pressure of $\approx 2.5 \times 10^8$ Pa), (2) at high ($\approx 95\%$) relative humidity ($\approx 8 \times 10^6$ Pa), (3) in *Ca-free NaCl buffer*, and (4) in *Ca05 NaCl buffer*. For some special cases, experiments were carried out with *Ca50 NaCl buffer*.

6.1.1 Influence on Inter-Membrane Interactions

As discussed in Sect. 2.1.2, the interactions between neighboring LPS membranes are result of a complex interplay of various generic interactions. Neutron scattering

Table 6.1 Lamellar spacings d of Lipid A and rough mutant LPS multilayers subject to various osmotic pressures and buffer conditions at $T = 60°C$

Environmental conditions	d of Lipid A [Å]	d of LPS Re [Å]	d of LPS Ra [Å]
Low relative humidity ($\approx 20\%$)	46.4	49.4	78.9
High relative humidity ($\approx 95\%$)	47.5	53.9	83.2
Ca-free NaCl buffer	53.5	Unstable	Unstable
Ca05 NaCl buffer	47.6	50.5	90.5
Ca50 NaCl buffer	46.9	50.2	90.4

from the samples subject to various relative humidities and under bulk buffers yields insight of the inter-membrane interactions at various osmotic pressures and ion concentrations. The lamellar periodicities d of Lipid A and LPS membrane multilayers, calculated from the q_z positions of the Bragg peaks (see Fig. 6.1 right) at various conditions, are summarized in Table 6.1. As a clear tendency, molecules with longer saccharide headgroups exhibit larger lamellar periodicity d under all experimental conditions, which shows a reasonable agreement with previous reports [1, 2]. Moreover, changes in the periodicity d between low and high humidity become more pronounced for the molecules with longer saccharide head groups ($\Delta d \approx 1$ Å for Lipid A, $\Delta d \approx 4.5$ Å for LPS Re and LPS Ra), reflecting their higher compressibility, since the longer and more complex headgroups possess a higher degree of conformational freedom.

It is notable that only Lipid A forms stable multilayers in calcium-free buffer, with a finite spacing of $d = 53.5$ Å. This implies that the attractive interactions (v. d. Waals, carbohydrate-mediated hydrogen bonds) between Lipid A membranes can overcome the electrostatic repulsion, since Lipid A membranes have the lowest density of negatively charged groups among the studied systems (see Sect. 3.1; Table 3.1).

On the other hand, LPS Re and LPS Ra membranes are not able to form stable lamellae in the absence of calcium ions, due to stronger electrostatic repulsion and due to the steric repulsion of the bulkier saccharide headgroups. This finding is consistent with a previous diffraction study on LPSs [1], where no periodic ordering was found for LPS Re and LPS Ra membranes in the absence of divalent cations.

Lipid A and both types of mutant LPSs form stable, well ordered multilayers in Ca05 NaCl buffer, which is in contrast to the significant weakening of inter-membrane confinement in Ca-free NaCl buffer. This shows that divalent calcium ions tighten the inter-membrane contacts. In fact, the lamellar spacings of Lipid A ($d = 47.6$ Å) and LPS Re ($d = 50.5$ Å) are almost identical or even smaller than those at $\approx 95\%$ relative humidity, which corresponds to an osmotic pressure of almost 10^7 Pa. This tendency becomes even more prominent at a high Ca^{2+} concentration (50 mM, in Ca50 NaCl buffer), where the spacings of Lipid A ($d = 46.9$ Å) and LPS Re ($d = 50.2$ Å) get close to those at 20% relative humidity, corresponding to an osmotic pressure of over 10^8 Pa. On the other hand, LPS Ra multilayers show a spacing of $d = 90$ Å in Ca5 NaCl buffer, which is significantly larger ($\Delta d \approx 7$ Å) than that at $\approx 95\%$ relative humidity.

This indicates that the longer and partially uncharged oligosaccharide moieties of LPS Ra resist to a certain extent the tightening of the inter-membrane contact by Ca^{2+} ions, due to their steric contribution.

6.1.2 Influence on Mechanical Properties

The mechanics of Lipid A and rough mutant LPS multilayers were investigated with a focus on two thermodynamic conditions: (1) at high humidity, where very strong scattering signals can be recorded, and (2) in Ca5 NaCl buffer, which provides the biologically most relevant environment for all the studied systems to form stable multilayers. As discussed in Sect. 5.1.3, the sample stability was verified from the symmetry of the Bragg sheets in Ω-direction. The multilayers

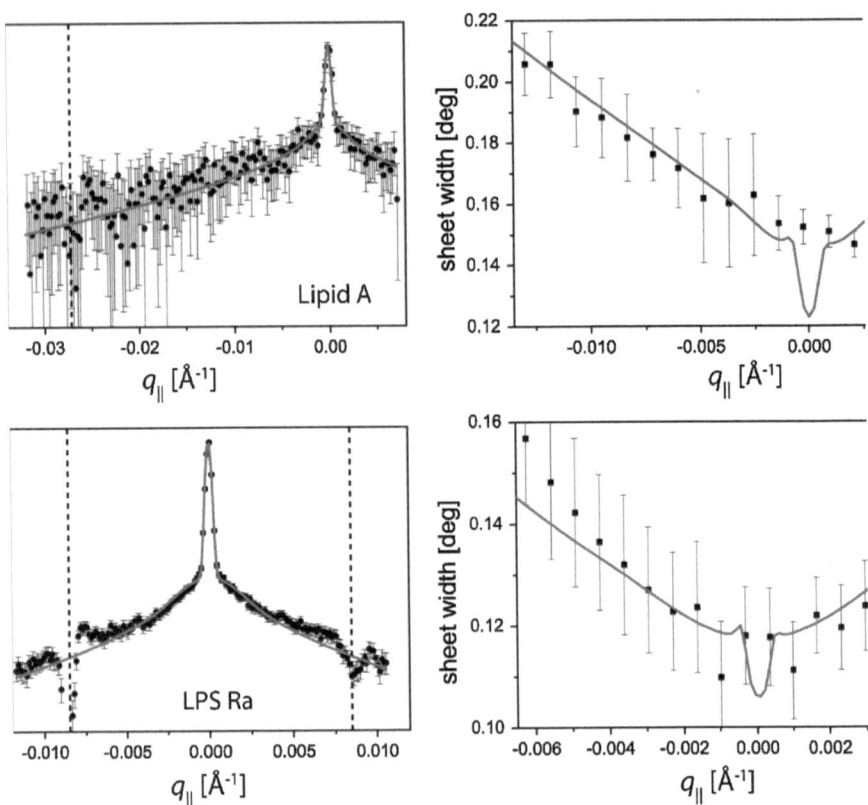

Fig. 6.2 Measured (*data points*) and simulated (*solid line*) second Bragg sheets of Lipid A (*top*) and LPS Ra (*bottom*) membrane multilayers at 60°C and $\approx 95\%$ relative humidity. (*Left column*) intensity integrated along Γ plotted as a function of $q_{\parallel}$. *Vertical dashed lines* indicate the positions of the sample horizons. (*Right column*) width of the sheet along Γ plotted as a function of $q_{\parallel}$

exhibited a high alignment with the planar substrate, as can be seen in Figs. 6.2 and 6.3 (left panels) from the narrow central specular maximum (angular width $\approx 0.15°$). As described in Sect. 4.1, the scattering signals were analyzed by varying the model parameters η, λ, and R in order to achieve the best match between simulations and experimental data. This allowed for the calculation of the compression modulus B and the bending modulus κ of each system. Figure 6.2 shows the integrated intensity (left) and the width (right) of the second Bragg sheet measured from Lipid A (top) and LPS Ra (bottom) membrane multilayers at 60°C and $\approx 95\%$ relative humidity. As motivated in Sect. 4.1, these representations enable a meaningful comparison between experimental data and simulations based on the free parameters η, λ, and R. Lipid A and LPS Ra both show pronounced second Bragg sheets. This can be attributed to the strong scattering length density contrast according to the uptake of D_2O by the head groups.

The parameters of the best matching model (solid lines in Fig. 6.2) are summarized in Table 6.2. The intensity of the second Bragg sheet of LPS Re (data not shown) is strongly suppressed due to the form factor corresponding to the scattering length density profile across the membranes, which prevents a quantitative analysis of the off-specular signals. The same problem occurs with Lipid A in *Ca-free NaCl buffer*, where the lamellar periodicity d is similar to that of LPS Re in high humidity (see Table 6.1).

The best matching values for the cut-off radius R (see Sect. 4.1) are at the order of 1 μm and similar to those found for synthetic glycolipids (see Chap. 5). First, the obtained bending rigidity of LPS Ra ($\kappa = 1.4$ k_BT) is more than two times higher than that of Lipid A ($\kappa = 0.6$ k_BT), indicating that the bulkier head groups significantly rigidify the membranes. On the other hand, both values are by an order of magnitude smaller than those obtained with DPPC and synthetic glycolipids with all-saturated C16 hydrocarbon chains (see Chap. 5). This finding can be attributed to the molecular structure of Lipid A and LPSs: bacterial lipids and LPSs have relatively short hydrocarbon chains (C10 and C12) and thus form bilayers with thinner hydrocarbon core. This can be assumed to result in significantly lower bending rigidities [3]. Furthermore, due to their higher packing parameter (i.e., the ratio between the cross-sectional areas occupied by hydrocarbon chains and saccharide head groups) [4], LPS membranes are expected to experience less steric strain while accommodating their head groups than phosphatidylcholine lipid membranes [5]. As a result, the bending rigidity of LPS membranes would be further reduced as there is only little contribution from the headgroups.

The obtained vertical cómpression moduli of Lipid A ($B = 9$ MPa) and LPS Ra ($B = 2.1$ MPa) seem to reflect the influence of the oligosaccharide head group structure on the inter-membrane potential. The observed tendency clearly indicates that the inter-membrane confinement of LPS Ra membranes is much softer than that of Lipid A membranes. This finding seems also plausible from the molecular structures, since the longer saccharide head groups of LPS Ra with their greater water uptake capability should possess a much higher compressional

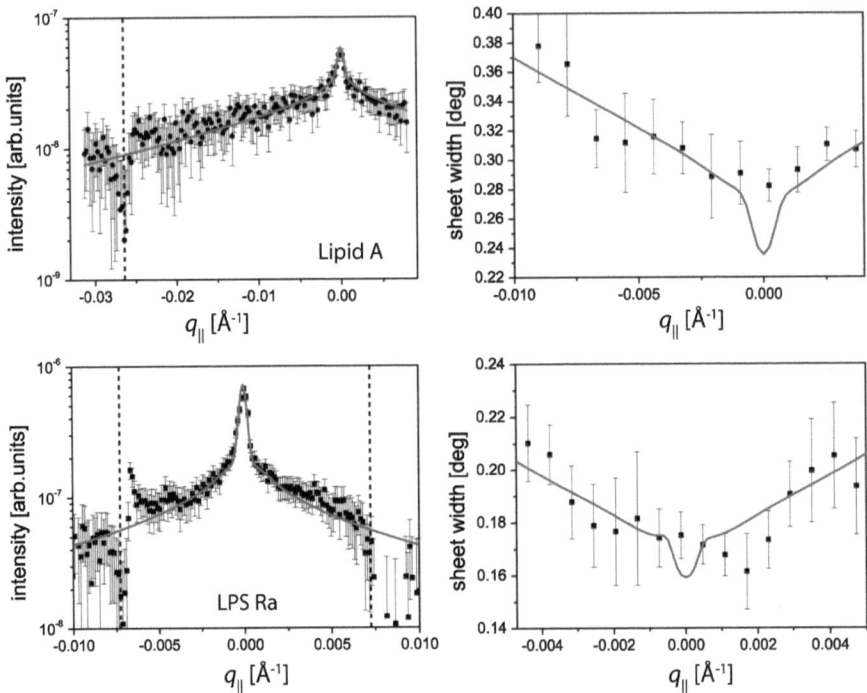

Fig. 6.3 Measured (*data points*) and simulated (*solid line*) second Bragg sheets of Lipid A (*top*) and LPS Ra (*bottom*) membrane multilayers at 60°C in *Ca05 NaCl buffer*. (*Left column*) intensity integrated along Γ plotted as a function of $q_{\parallel}$. *Vertical dashed lines* indicate the positions of the sample horizons. (*Right column*) width of the sheet along Γ plotted as a function of $q_{\parallel}$

flexibility than the compact head groups of Lipid A. Figure 6.3 shows the scattering signals of Lipid A (top) and LPS Ra (bottom) in *Ca05 NaCl buffer*. The parameters corresponding to the best matching model signals (solid lines) are summarized in Table 6.3. The bending rigidity of Lipid A ($\kappa = 0.8\ k_B T$) is about two times lower than that of LPS Ra ($\kappa = 1.7\ k_B T$), which follows the tendency observed at high humidity ($\kappa = 0.6\ k_B T$ and $\kappa = 1.4\ k_B T$, respectively). Moreover, the bending rigidity values are very similar to the corresponding values at high humidity. This suggests that the lateral interactions of the saccharide head groups in Ca^{2+}-loaded buffer are similar to those in the absence of liquid water. However, it is not possible to extract the bending rigidity in calcium-free buffer due to the loss of ordering of LPS Ra and due to the before mentioned suppression of the second Bragg sheet in the case of Lipid A. On the other hand, the obtained compression moduli in Ca5 NaCl buffer show clear differences from those at high humidity. As presented in Table 6.3, the values in bulk buffer ($B = 5$ MPa for Lipid A and $B = 1.1$ MPa for LPS Ra) are approximately by a factor of two smaller than those at high humidity (see Table 6.2).

Table 6.2 Parameters of the best matching models for Lipid A and LPS Ra membrane multi-layers at $T = 60°C$ and $\approx 95\%$ relative humidity

Molecule	d [Å]	η	λ [Å]	R [µm]	κ [$k_B T$]	B [MPa]
Lipid A	47.5	0.135	2.5	0.8	0.6	9
LPS Ra	83.2	0.08	6	0.8	1.4	2.1

Table 6.3 Parameters corresponding to the best matching models for Lipid A and LPS Ra membrane multilayers at $T = 60°C$ in *Ca05 NaCl buffer*

Molecule	d [Å]	η	λ [Å]	R [µm]	κ [$k_B T$]	B [MPa]
Lipid A	47.6	0.17	4	1.0	0.8	5
LPS Ra	90.5	0.09	9	1.2	1.7	1.1

6.1.3 Summary of Sect. 6.1

Solid-supported multilayers of Lipid A and rough mutant LPS membranes were shown to constitute a well-defined model system of interacting bacteria membranes, which can investigated by specular and off-specular neutron scattering for the quantitative determination of mechanical parameters. With increasing complexity of the studied molecules, clear tendencies were found in the mechanical properties: Longer and more flexible saccharide head groups coincide with a significantly softer inter-membrane confinement but at the same time with a higher membrane bending rigidity. Generally, the bending rigidities of Lipid A and mutant LPS membranes were found to be much lower than those of commonly studied phospholipid membranes. The compression modulus of LPS Ra membranes in Ca5 NaCl buffer was found to be comparable to that of interacting Gentiobiose lipid membranes with neutral gentiobiose head groups in pure water ($B = 0.9$ MPa, see Sect. 5.1.3), which also swell by about 7 Å upon addition of bulk water. However, in that case the decrease in the compression modulus is much more significant (by a factor of 20). This indicates that the presence of Ca^{2+} between the charged carbohydrates of Lipid A and mutant LPS molecules enhances the inter-membrane confinement, which is consistent with the fact that these systems are stabilized in the presence of Ca^{2+}.

6.2 Influence of Divalent Cations on the Conformation of Wild-Type Lipopolysaccharides

To investigate the influence of divalent cations on the conformation of the charged saccharides at the outer surface of Gram-negative bacteria, a single solid-supported monolayer (see Sect. 2.1.4.2) of lipopolysaccharides purified from *Pseudomonas aeruginosa* strain dps 89 was created (see Sects. 3.1.2 and 3.2.4).

Fig. 6.4 Sketch of the
studied solid-supported LPS
monolayer under bulk buffer.
A hydrocarbon chain layer,
B dense core saccharide
layer, *C* sparse O-side chain
region

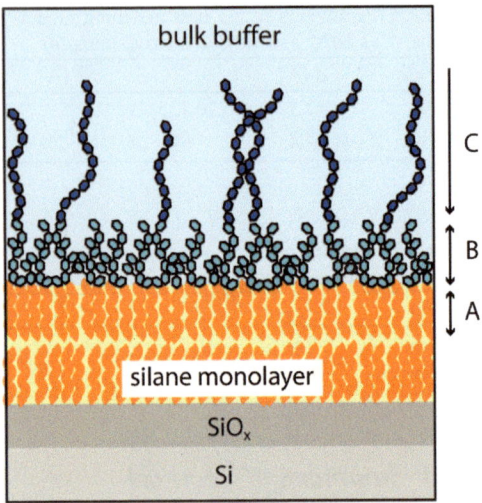

The LPS monolayer (Fig. 6.4) was deposited by vesicle fusion onto a silicon
substrate coated with an alkylsilane monolayer which mimics the inner leaflet of
the outer bacteria membrane and studied by high-energy specular X-ray reflectivity. The use of high X-ray energies (here 22 keV) was established only recently
[6, 7] and allows for the high-resolution structural analysis of thin films at the
solid/liquid interface. The measured reflectivity signals were modeled using analytically parameterized electron density profiles, which were compared with
Monte Carlo (MC) simulations based on a minimal computer model of LPS surfaces. Details of the experimental setup are presented in Sect. 3.3.1.1.

6.2.1 Effect of Divalent Cations: Electron Density Profile of LPS Monolayers

Figure 6.5 (left panel) presents the reflectivity curves $R \times q_z^4$ measured in *Ca-free
KCl buffer* (open circles) and *Ca50 KCl buffer* (filled circles), showing a clear
difference in the global shape of the reflectivity curves.[1] The measured reflectivity
signals from the stratified system (Fig. 6.4) were represented with the following
simplified model: First, the silicon substrate was treated as a semi-infinite medium
with a constant electron density $\rho_{Si} = 0.713$ e$^-$/Å3. For simplicity, the native
silicon oxide layer that has almost identical electron density was considered as part

---

[1] Here, 50 mM CaCl$_2$ was used for the direct comparison to the MC simulations where a high
calcium concentration was assumed. However, regarding the only small differences in structure
and mechanical properties of interacting mutant LPS membrane multilayers found with 5 mM
and 50 mM CaCl$_2$ (see Sect. 6.1), much lower calcium concentrations are likely sufficient to
induce the conformational changes observed here.

of the silicon medium. The hydrocarbon chain region, composed of the substrate-bound ODTMS layer and the LPS hydrocarbon chain layer, was considered as one single slab with constant electron density ρ_{HC} and thickness d_{HC}. The saccharide headgroups were represented by two sections; one slab with constant electron density ρ_S and thickness d_S, and the other section where the electron density continuously decays towards that of bulk buffer $\rho_B = \rho_{H_2O} = 0.336$ e$^-$/Å3. The former corresponds to the compact invariant part constituted by the core saccharides, while the latter represents the sparse and polydisperse O-sidechains. The roughnesses at the silicon/hydrocarbon and hydrocarbon/core saccharide interfaces were accounted for by modeling the electron density gradients as error functions, characterized by the RMS roughness $\sigma_{Si \to HC}$ and $\sigma_{HC \to S}$, respectively (see Sect. 2.2.2). However, since the O-sidechains are expressed at a low surface density ($\approx 10\%$) and polydisperse (see Sect. 3.1.2), a slab model definition is not suited to describe the electron density profile of the O-sidechains in buffer. Instead, a continuous decay of the electron density towards bulk buffer was assumed and modeled using a stretched exponential decay function, characterized by the decay length Λ, and the stretching exponent h. To achieve a common parameter set for $z < 0$, the thickness and electron density of the hydrocarbon region (ρ_{HC}, and d_{HC}, respectively) as well as the roughness of the Si/hydrocarbon interface ($\sigma_{Si \to HC}$) were simultaneously fitted to the reflectivity signals measured in the presence and absence of Ca^{2+} ions. This simplification is justified as the reflectivity measurements were successively carried out with the same sample in *Ca-free KCl buffer* and *Ca50 KCl buffer*, and reduces the number of fitting parameters. In the left panel of Fig. 6.5, the best fits are superimposed to the reflectivity curves in *Ca-free KCl buffer* and *Ca50 KCl buffer*, and corresponding model parameters are summarized in Table 6.4. The electron density profiles reconstructed from the models

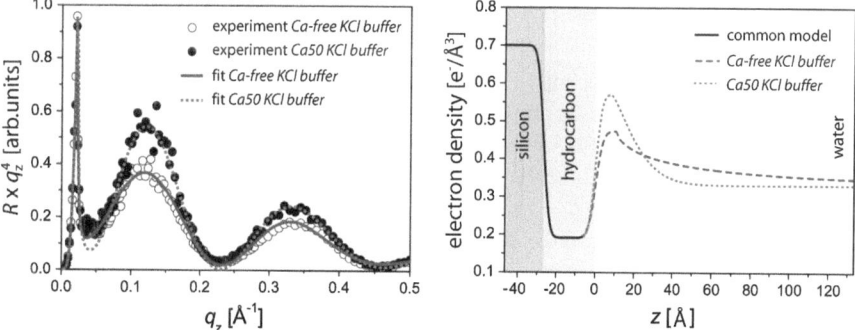

Fig. 6.5 (*Left*) Reflectivity *curves* of solid-supported LPS monolayer in *Ca-free KCl buffer* (*open circles*) and *Ca50 KCl buffer* (*filled circles*). *Solid* and *dashed lines* represent the fits corresponding to the best-matching model parameters. (*Right*) Electron density profiles corresponding to the best-matching model parameters. The interface between hydrocarbon slab and core saccharide slab coincides with $z = 0$. The core saccharides and O-side chains exhibit a significant difference in the conformation in the absence (*dashed line*) and presence (*dotted line*) of Ca^{2+}. *Solid line* common model

Table 6.4 Best matching model parameters used to represent the reflectivity curves measured in the presence and absence of 50 mM CaCl$_2$

Common parameters	Ca-free KCl buffer and Ca50 KCl buffer	
$\sigma_{Si \to HC}$ [Å]	2.3	
ρ_{HC} [e$^-$/Å^3]	0.19	
d_{HC} [Å]	26.9	
Free parameters	Ca-free KCl buffer	Ca50 KCl buffer
$\sigma_{HC \to S}$ [Å]	2.9	2.6
ρ_S [e$^-$/Å^3]	0.47	0.57
d_S [Å]	10.7	7.8
Λ [Å]	34.4	15.6
h	0.30	0.68

are shown in the right panel of Fig. 6.5. Although the five free fitting parameters imply a certain ambiguity, only the presented model fulfills the requirement of electron number conservation. Other models with a comparably low fit mean square deviation (MSD) had to be discarded, as the integrated electron density was significantly altered by the presence of calcium ions.

The obtained thickness ($d_{HC} = 26.9$ Å) of the "unified" hydrocarbon slab (chains from LPS and silane) seems very reasonable regarding the length of the ODTMS molecules and that of the LPS hydrocarbon chains. The electron density of this slab, $\rho_{HC} = 0.19$ e$^-$/Å^3, is smaller than the value previously reported for rough mutant LPS monolayers ($\rho \cong 0.3$ e$^-$/Å^3) at the air/water interface [8]. This can be attributed to the lower lateral density of chains in ODTMS and LPS monolayers, and to the local density minimum between two opposing hydrocarbon chains (called "methyl dip"), which was not explicitly included in the model in order to minimize the number of parameters.

The reconstructed electron density profiles (Fig. 6.5 right) indicate that Ca^{2+} ions induce two prominent differences. First, the electron density in the core saccharide slab is significantly higher ($\approx 20\%$) in the presence of Ca^{2+} ions ($\rho_S = 0.57$ e$^-$/Å^3) than in the absence ($\rho_S = 0.47$ e$^-$/Å^3). Second, in the presence of Ca^{2+}, the electron density of the O-sidechain region decays much steeper to the level of bulk buffer (at $z > 10$ Å), which corresponds to an increase in the stretching exponent (from $h = 0.30$ in the absence to $h = 0.68$ in the presence of Ca^{2+} ions) and a decrease in the decay length (from $\Lambda = 34.4$ Å in the absence to $\Lambda = 15.6$ Å in the presence of Ca^{2+} ions).

6.2.2 Modeling of LPS Saccharide Conformation by Coarse-Grained Monte Carlo Simulations

To simulate the conformation of LPS molecules in the absence and presence of Ca^{2+} ions, coarse-grained MC simulations were carried out by D. A. Pink and B. E. Quinn

(St Francis Xavier University, Antigonish, Canada). The simulation volume contained 100 LPS molecules based on a "minimal computer model" of lipopoly-saccharides. It was assumed that 10% of the LPS molecules possess an O-sidechain, while 90% of the molecules (including 70% rough and 20% semi-rough LPSs) are all rough LPSs (see Sect. 3.1.2). Accordingly, 10 randomly chosen LPS molecules possessed O-sidechains. In order to minimize the simulation volume the length of the O-sidechains was set to be 20 repeat units (60 saccharides), which is sufficient to describe natural B-band polysaccharides with 20–50 repeating units (see Sect. 3.1.2).The bulk densities of monovalent and divalent ions in the simulation volume were consistent with the experimental buffer conditions (100 mM KCl and 0 mM or 50 mM $CaCl_2$). Details of the MC simulations are given elsewhere [8–10]. Figure 6.6 shows instantaneous pictures of the simulation volume after equilibration in the absence (a) and in the presence (b) of divalent cations. The pictures show a significant change in the conformation of LPS molecules: the long O-sidechains collapse towards the core saccharide region in the presence of divalent cations. This effect was quantified by calculating the time-averaged vertical number density profiles of the saccharide moieties (Fig. 6.6c). The profiles show that divalent cat-ions induce a clear increase of the saccharide density in the core saccharide region, while the tail of the distribution decays to zero at much lower z in the presence of divalent cations. It should be noted that the simulations showed no clear sign of aggregation of the O-sidechains in the absence of divalent cations, despite of significant electrostatic screening. In contrast, in the presence of divalent cations, the O-sidechains are bound together via dynamic bridging by divalent cations (Fig. 6.6b). As a result, the O-sidechains undergo a massive collapse towards the surface of dense core saccharides. Previous, though less realistic, computer

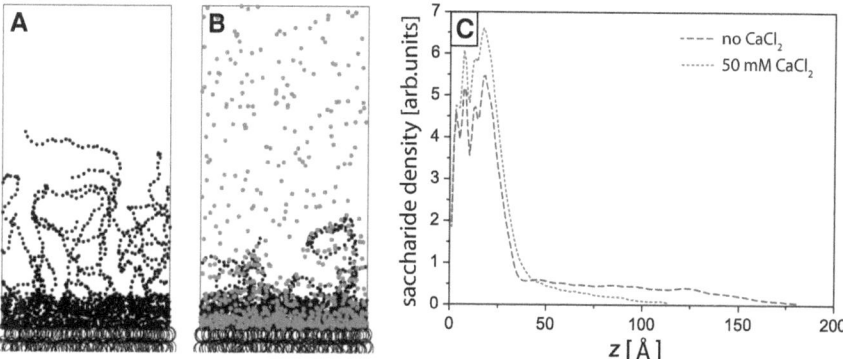

Fig. 6.6 Instantaneous pictures of the MC simulation after equilibration in the absence (**a**) and presence (**b**) of divalent cations (Ca^{2+}). The simulations were carried out by D.A. Pink and B.E. Quinn. Hydrocarbon moieties are indicated by *open large circles*, saccharide groups by *small filled black circles* and divalent cations by *small filled grey circles*. Monovalent ions (K^+ and Cl^-) are not shown. (**c**) Time-averaged number density profiles of saccharide groups in the aqueous region ($z > 0$) calculated from the MC simulations in the absence (*broken line*) and presence (*dashed line*) of divalent cations

Fig. 6.7 Time-averaged electron density profiles in the aqueous region ($z > 0$) calculated from the MC simulations in the absence (*broken line*) and presence (*dashed line*) of divalent cations. The simulations were carried out by D.A. Pink and B.E. Quinn

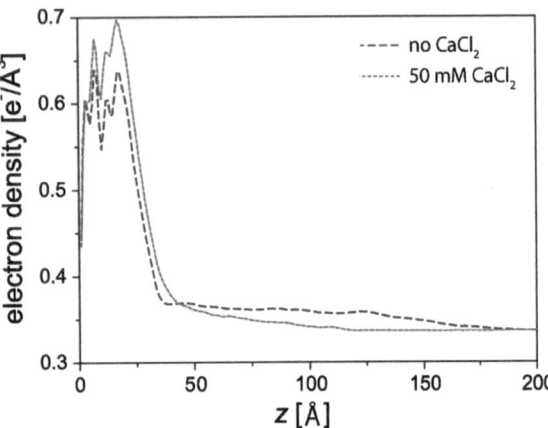

simulations showing similar conformational changes, suggested that these changes result in a barrier against the intrusion of antimicrobial peptides into the membrane core [9], but were not yet verified by experiments.

For the direct comparison to the experimental results presented in Fig. 6.5, time-averaged electron density profiles were calculated from the MC simulations in the absence and in the presence of divalent cations (Fig. 6.7). The calculated profiles show the same tendency as those reconstructed from the measured reflectivity curves (Fig. 6.5 right): Divalent cations induce a significant increase in electron density in the core saccharide region and a steep decay of the electron density to the bulk level. The differences in absolute electron density values and the detailed shape of the electron density profiles between experiments and computer simulations can be attributed to the uncertainty ($\approx 15\%$) in the lateral density of the LPS monolayer as well as to the lack of semi-rough LPSs (see Sect. 3.1.2) in the computer simulation.

6.2.3 Summary of Sect. 6.2

A realistic model of outer bacteria membranes was created by the deposition of a monolayer of purified lipopolysaccharides onto a planar substrate functionalized with a hydrophobic alkyl silane monolayer. This model system was structurally characterized using high-energy specular X-ray reflectometry in bulk buffer solution. The influence of divalent calcium ions on the conformation of the charged LPS O-sidechains was investigated. The electron density profiles of the monolayer, reconstructed from the reflectivity measurements in the absence and presence of Ca^{2+} ions, indicated that calcium induces a collapse of the O-sidechains towards the core saccharide region. The same tendency was observed in MC simulations carried out by collaboration partners, where the simulated system represented the experimental sample composition. The obtained results constitute the first experimental demonstration that divalent cations induce the conformational changes which were previously suggested to lead to a barrier against the intrusion of cationic antimicrobial peptides into the outer membranes of Gram-negative bacteria.

6.3 Concentration Profiles of Monovalent and Divalent Cations at Bacteria Surfaces

As motivated in Chap.1, the molecular conformation of LPS surfaces and the resistance of Gram-negative bacteria against antimicrobial peptides are intimately linked to the distribution of ions. Thus, it can be considered eminently important to gain information not only on the LPS conformation under various conditions (see Sect. 6.2), but also on the associated ion distributions. Grazing incidence X-ray fluorescence (GIXF) allows for the localisation of chemical elements near an interface. The sensitivity of the method has been recently exploited to resolve the weak depletion of ions in the vicinity of the water surface [11]. To date, GIXF has been established for solid-supported thin films [12] and surfactant monolayers at the air/water interface [13, 14]. If detailed information on the electronic structure of the system is available, the method can be readily extended to localize elements in layered organic systems [15]. However, the evanescent X-ray field required for GIXF measurements can only be established at the air/water interface, which renders solid-supported model systems inadequate.[2] Instead, the model of a bacteria surface was established by the preparation of an LPS Re monolayer at the air/water interface (Fig. 6.8) using a Langmuir trough (see Sects. 3.1.2 and 3.2.1.3). This approach, where macroscopic areas of oriented monolayers can be studied (see Sect. 2.2.1.2), has been widely used to investigate phospholipid systems. Among many other techniques, X-ray scattering has been shown to have a large potential to examine the monolayer structure with great detail [8, 16–18]. Here, the LPS Re monolayers at the air/water interface were studied by grazing incidence X-ray scattering and X-ray fluorescence to examine monolayer structure and ion distributions in the absence and presence of divalent cations.

6.3.1 Influence of Divalent Cations on Molecular Interactions in Langmuir Monolayers of LPS Re

Prior to the X-ray experiments, pressure-area isotherms were recorded. Figure 6.8 (right) shows the pressure-area isotherms of LPS Re monolayers on *Ca-free NaCl*

---

[2] For X-rays, total reflection can only occur if the beam travels from a medium with lower electron density to a medium with higher electron density (see Sect. 2.2). In contrast to all commonly used solid substrate materials, the gas phase has a lower electron density than water.

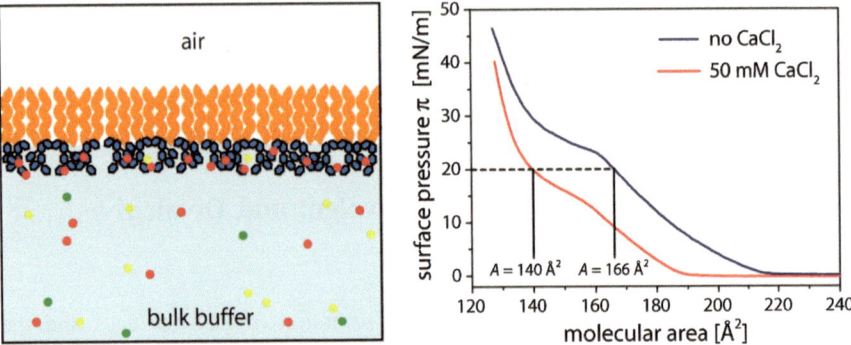

Fig. 6.8 (*Left*) Sketch of the studied LPS Re monolayer at the air/water interface. Ions of various types are indicated with dots of various colors. (*Right*) Pressure-area isotherms of LPS Re monolayers on *Ca-free NaCl buffer* (*blue*) and *Ca50 NaCl buffer* (*red*). X-ray experiments were carried out at a surface pressure of $\pi = 20$ mN/m (*horizontal dashed line*), where the areas per molecule are $A \approx 166$ Å^2 and $A \approx 140$ Å^2 in the absence and in the presence of calcium, respectively

buffer (blue) and *Ca50 NaCl buffer* (red) at 20°C. In the presence of calcium, the onset of pressure increase appears at $A \approx 185$ Å^2. This value is smaller than that obtained in the absence of calcium ($A \approx 215$ Å^2), suggesting that Ca^{2+} ions decrease the range of lateral repulsive interactions of the headgroups. On the other hand, when the monolayer is compressed to a liquid-condensed phase ($A < 140$ Å^2), the lateral compressibilities (proportional to the slope of the isotherms) on both buffers become comparable, implying that here the area per molecule of the monolayer is governed by the hard repulsion of hydrocarbon chains. The plateau-like regimes observed on both buffers can be interpreted as the coexistence of fluid (liquid-expanded) and solid (liquid-condensed) phases (see Sect. 2.1.4.1). The shoulder regime appeared at $\pi = 25$–30 mN/m in the absence of calcium, while the corresponding feature is observed at much lower surface pressure ($\pi = 15 \approx 20$ mN/m) in the presence of calcium. The observed differences in the isotherms suggest that lateral interactions between LPS molecules are significantly altered by Ca^{2+} ions.

6.3.2 Influence of Divalent Cations on Electron Density Profiles of LPS Re Monolayers

The LPS Re monolayers were studied at 20°C and $\pi = 20$ mN/m, a lateral pressure which reproduces conditions equivalent to those in live bacterial membranes. According to the isotherms (Fig. 6.8) this corresponds to molecular areas of about 166 Å^2 and 140 Å^2 on *Ca-free NaCl buffer* and on *Ca50 NaCl*

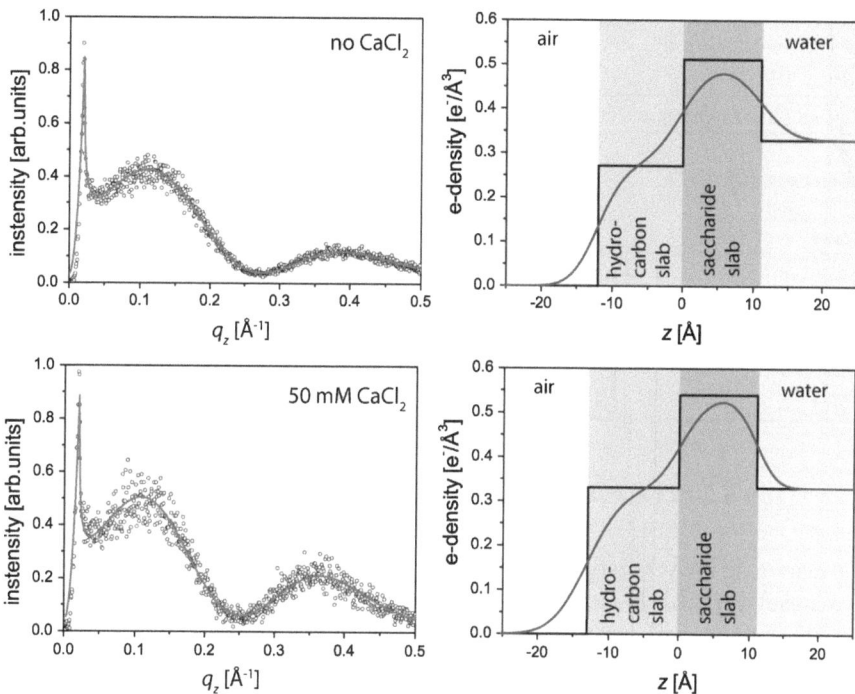

Fig. 6.9 GIXOS signals and reconstructed electron density profile of LPS Re monolayers at a surface pressure of $\pi = 20$ mN/m on *Ca-free NaCl buffer* (*top*) and on *Ca50 NaCl buffer* (*bottom*). (*Left column*) data points from GIXOS experiments (*open circles*) and best matching models (*solid lines*). (*Right column*) reconstructed electron density profiles (*solid curve*) and underlying slab model (*black straight lines*)

buffer, respectively. In the first step the electron density profiles of the monolayers in the presence and absence of calcium was characterized. To obtain vertical electron density profiles, GIXOS measurements were carried out (see Sect. 3.1.2). As described in previous studies [8, 19], a two-slab model including interfacial roughness (see Sect. 2.2.1.2) was used to fit the measured GIXOS curves: the first slab represents hydrocarbon chains and the second one the saccharide headgroups.

Measured GIXOS signals and modeled signals corresponding to the best fitting parameters are shown in Fig. 6.9 (left panels). The corresponding electron density profiles and slab models are shown in the right panels of the figure. The model parameters are summarized in Table 6.5. The obtained electronic structures are in good agreement with previous structural studies on LPS mutant membranes [1, 8, 20–22]. The generally lower electron densities found on *Ca-free NaCl buffer* seem to reflect the larger average molecular area compared to that on *Ca50 NaCl buffer* (see Fig. 6.8).

Table 6.5 Best matching model parameters used to represent the GIXOS signals measured with LPS Re monolayers at a surface pressure of $\pi = 20$ mN/m on *Ca-free NaCl buffer* and *Ca50 NaCl buffer*

Ca-free NaCl buffer, $A = 166$ Å^2	ρ [e$^-$/Å^3]	d [Å]	σ [Å]
Air	0.00	N/A	N/A
Hydrocarbon chains	0.27	12	2.9
Head group	0.51	11	3.8
Water	0.33	N/A	3.6
Ca50 NaCl buffer, $A = 140$ Å^2	ρ [e$^-$/Å^3]	d [Å]	σ [Å]
Air	0.00	N/A	N/A
Hydrocarbon chains	0.33	13	3.9
Head group	0.54	11	3.2
Water	0.33	N/A	2.4

6.3.3 Ion Concentration Profiles at LPS Re Monolayers

To determine the vertical profiles of monovalent and divalent cations near the LPS Re monolayers, GIXF measurements were carried out. This provides information on an absolute scale on the ionic concentration profile normal to the interface. Together with the known area per molecule, the number of ions enriched per LPS Re molecule can be quantitatively "counted". In order to fully explore this potential, the fluorescence signals were theoretically modeled. The taken approach for the simulation and interpretation of GIXF signals is developed and discussed in detail in Sect. 4.3. For the calculations, the slab models obtained from GIXOS experiments were used.[3] As motivated in Sect. 4.3, interfacial roughness does not have to be considered. The difference between the monovalent salts used for the two techniques (see Sect. 3.1.2) is negligible.[4] For the blank buffer a constant ion concentration was assumed ($c_0 = 0.1$ M for K$^+$, $c_0 = 0.05$ M for Ca^{2+}) starting at the first interface (i.e., the air/water interface), since the influence of ion depletion near the water surface [11] was negligibly smaller than the significant

---

[3] The LPS Re molecules used for GIXOS experiments on the one hand and for X-ray fluorescence measurements on the other hand originate from different biological sources (from the strain F515 of *E. coli.* and from the strain R595 of Salmonella enterica sv. Minnesota, respectively, see Sect. 3.1.2). However, the ensuing small deviations of the electron density profile employed for X-ray fluorescence analysis have no significant influence on the data interpretation.

[4] The number of electrons per LPS Re molecule in the headgroup slab ($N = \rho A d$) is at the order of 1,000, while the numbers of electrons per K$^+$ and Na$^+$ differ only by 8. Considering the number of excess K$^+$ ions per LPS Re molecule ($N = 2.32$) observed on *Ca-free KCl buffer*, the difference in electron density resulting from the different choice of monovalent salt can be neglected (<2%). On *Ca50 KCl buffer*, where there is only a very slight enrichment of monovalent ions in the headgroup slab, the effect is even weaker.

accumulation of ions in the presence of LPS monolayers. The excess concentration profile of cations condensed near the charged LPS Re monolayer was parameterized as:

$$c_{ex}(z) \propto c_{max} \cdot z \cdot \exp\left(-z^2/2z_{max}^2\right),$$

with a vanishing ion density at the chain/carbohydrate interface, $c_{ex}(z = 0) = 0$, and the requirement $lim_{z\to\infty}c_{ex}(z) = 0$. This allows modeling ion distributions that possess a concentration maximum with a smooth decay to the bulk concentration with only two free parameters: 1) the concentration maximum c_{max} and 2) the z-position z_{max} of this maximum. Figure 6.10 shows the measured K^+ fluorescence signals from *Ca-free KCl buffer* without (squares) and with (triangles) an LPS Re monolayer as a function of q_z. For better visibility the two curves are offset by one order of magnitude. The modeled signals (solid lines) are superimposed to the experimental data points. X-ray absorption by water was accounted for by using the absorption coefficient of pure water at the used X-ray energy (8 keV), corrected for the bulk ion concentrations. The effect of fluorescence re-absorption is small compared to that of the illumination absorption in the given q_z-range, and was thus ignored in the simulations (see Sect. 4.3.2.1). Obviously the global shape of the experimental curves is well captured by the modeled curves over several orders of magnitude in intensity. The remaining deviations can be attributed to non-considered geometrical effects. Therefore, the following interpretation is based on the buffer-normalized signals (fluorescence signals recorded with LPS monolayer are divided by the corresponding signals recorded without monolayer). In this way possible error sources originating from geometrical corrections and absorption effects are avoided (see Sect. 4.3.2.1).

Figure 6.11 (left panel) shows the measured buffer-normalized K^+ fluorescence signal (open circles) from an LPS Re monolayer on *Ca-free KCl buffer*. Some qualitative conclusions can be drawn immediately: The data points are found significantly above unity for q_z-values below the critical angle of total reflection.

Fig. 6.10 Fluorescence intensity as a function of q_z. Experimental (*open symbols*) and modeled (*solid lines*) K^+ fluorescence from an LPS Re monolayer (*triangles*) and from a blank buffer (*squares*). For better visibility the two *curves* are shifted vertically in the plot. The q_z value corresponding to the critical angle of total reflection (q_z^c) is indicated with a vertical *dashed line*

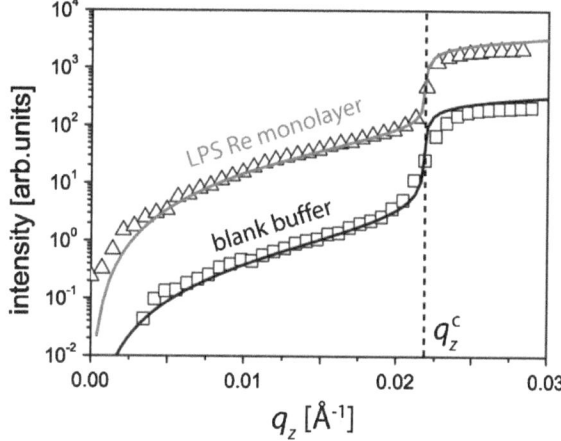

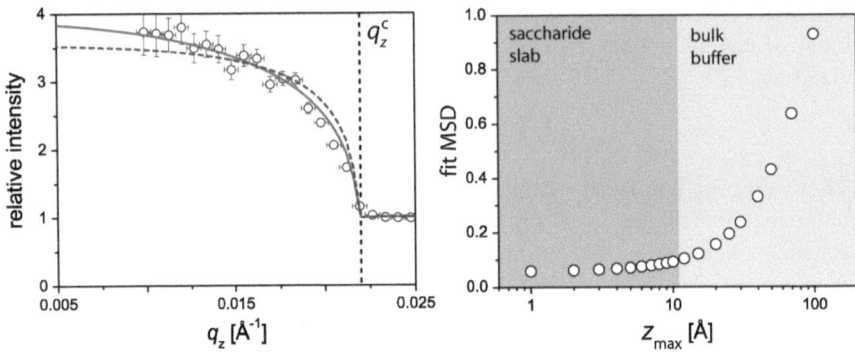

Fig. 6.11 (*Left*): Buffer-normalized K^+ fluorescence intensity on Ca-free KCl buffer as a function of q_z. Measured data points (*open circles*) and best matching model signals for $z_{max} = 5$ Å (*solid line*) and $z_{max} = 20$ Å (*dashed line*). The q_z value corresponding to the critical angle of total reflection (q_z^c) is indicated with a *vertical dashed line*. *Vertical error bars* were obtained via Gaussian error propagation from the parameters used to fit the fluorescence spectra. *Horizontal error bars* reflect the finite angular resolution. (*Right*): Mean square deviation (MSD) of the fit as a function of z_{max}

This indicates a surface-near enrichment of K^+ ions compared to the blank buffer (without LPS monolayer). The signal was fitted with the two free parameters c_{max} and z_{max}. For this purpose, z_{max} was varied stepwise while a least-square fit of the peak concentration c_{max} was performed for each step. This is shown in Fig. 6.11 (right panel), where the mean square deviation (MSD) of the fit is plotted as a function of z_{max}. Best accordance is found for small values of z_{max} ($z_{max} < 10$ Å), while for higher z_{max} values the MSD starts increasing significantly. Regarding the saccharide headgroup thickness of 11 Å (see Table 6.5), it can be concluded that the K^+ concentration peak is localized within the headgroup region of the LPS Re monolayer. For comparison, the best fits achieved with $z_{max} = 5$ Å (center of the saccharide slab) and with $z_{max} = 20$ Å (outside the saccharide slab) are superimposed to the experimental data points in Fig. 6.11 (left panel). Obviously, $z_{max} = 20$ Å is incompatible with the measured signals, while for $z_{max} = 5$ Å the data points are well represented by the modeled signal. In fact, the same model was used to reproduce the non-normalized K^+ fluorescence signals in Fig. 6.10. For $z_{max} = 5$ Å, best accordance is found for $c_{max} \approx 3$ M, corresponding to the solid blue line in Fig. 6.11 (left). The integrated K^+ excess concentration corresponds to a number of about 2.3 K^+ ions per LPS Re molecule. Although this value depends on the exact choice of z_{max}, it changes only by a few percent within the determined z_{max} range ($z_{max} < 10$ Å).

Figure 6.12 shows the measured buffer-normalized K^+ (open circles) and Ca^{2+} (filled circles) fluorescence signals from an LPS Re monolayer on *Ca50 KCl buffer*. The Ca^{2+} fluorescence signal is significantly elevated above unity below q_z^c, indicating a strong enrichment of Ca^{2+} ions near the LPS Re surface. The Ca^{2+} data points are well represented by a concentration profile with a maximum at

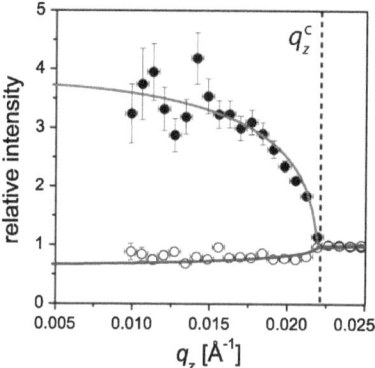

Fig. 6.12 Buffer-normalized K⁺ (*open circles*) and Ca²⁺ (*filled circles*) fluorescence intensity on *Ca50 KCl buffer* as a function of q_z. The best matching Ca²⁺ model signal for $z_{max} = 5$ Å is indicated with a *solid line*. The *solid line* superimposed to the K⁺ data points corresponds to a constant K⁺ concentration starting from the chain/saccharide interface. The q_z value corresponding to the critical angle of total reflection (q_z^c) is indicated with a *vertical dashed line*. Vertical *error bars* were obtained via Gaussian error propagation from the parameters used to fit the fluorescence spectra. *Horizontal error bars* reflect the finite angular resolution

$z_{max} = 5$ Å. The corresponding modeled Ca²⁺ signal (solid red line) is superimposed to the data points. Here, a best match is found for $c_{max} \approx 1.5$ M, coinciding with about 1.05 Ca²⁺ ions per LPS Re molecule.

Below the critical angle of total reflection, the buffer-normalized K⁺ fluorescence signal is significantly lower in the presence (Fig. 6.12) than in the absence (Fig. 6.11) of calcium, which clearly demonstrates that monovalent K⁺ ions are displaced by the divalent Ca²⁺ ions near the LPS monolayer. Moreover, below K⁺ data points below q_z^c even fall below unity in the presence of calcium. This appears to suggest a surface-near depletion of K⁺ below the bulk concentration, but great care must be taken while interpreting this effect. For a correct conclusion, quantitative modeling of the fluorescence signal is essential. In fact, the K⁺ signal on *Ca50 KCl buffer* (solid blue line in Fig. 6.12) was modeled with a constant K⁺ concentration profile ($c_0 = 0.1$ M) starting from the chain/saccharide interface. Obviously the corresponding modeled signal shows the same trend as the experimental data points with a reasonable match within the experimental error. Accordingly, there are two causes for the weaker K⁺ fluorescence signal from the monolayer system compared to the blank buffer below q_z^c:

(1) There are no ions within the alkyl chain layer, suppressing the fluorescence from the first illuminated 13 Å.
(2) To a lesser extent, the higher electron density of the saccharide region compared to water (see Table 6.5) reduces the effective penetration depth of the evanescent field into the bulk buffer and hence the illuminated volume.

In this light, it can be concluded that neither significant enrichment nor depletion of K⁺ near the monolayer in presence of 50 mM CaCl₂ is found within the given accuracy. On the other hand, any substantial contribution of K^+ to the compensation of the negative charge of the LPS Re monolayer can be excluded. Gouy-Chapman theory [23–25] qualitatively predicts the same tendency in an analytical continuum description of ion distributions near charged surfaces (see Sect. 4.2). Another important conclusion of this description is that for large enough length scales the charge of the interface is completely compensated.[5]

Because the contribution of Cl^- to the charge compensation at the interface is negligibly small, the surface charge compensation is dominated by K^+ in the absence of Ca^{2+} ions, whereas it is vastly dominated by Ca^{2+} ions if they are present in sufficient amount. This indicates that the effective charge of LPS Re is $Q \approx -2.3e$ in the absence and $Q \approx -2.1e$ in the presence of 50 mM $CaCl_2$, respectively. If one considers the uncertainty in the area per molecule ($\sim 10\%$), the net charge per LPS Re molecule predicted from two independent subphase conditions seem to achieve good quantitative agreement.

The charge of LPS molecules at environmental buffer conditions is still under debate [26]. Hagge et al. [27]. deduced the charge per LPS Re molecule, $Q \approx -3.6e$, from the average molecular structure. However, it is difficult to determine and predict the degree of ionization of lipids on membrane surfaces solely from molecular structures, since the deprotonation of the phosphate and carboxyl groups depends via the local pH on the electric potential at the surface [28, 29]. Generally, the remaining polydispersity of purified biological materials can limit the accuracy with which the average number of chargeable groups per molecule is known. In this thesis, a strategy for the precise determination of the actual charge density of biological surfaces under environmental conditions is shown.

6.3.4 Modeling of Ion Concentration Profiles at LPS Re Monolayers by Coarse-Grained Monte Carlo Simulations

To simulate the concentration profiles of monovalent and divalent ions near the LPS Re monolayers, coarse-grained MC simulations were carried out by D. A. Pink and B. E. Quinn (St Francis Xavier University, Antigonish, Canada). The simulation volume contained 100 LPS Re molecules based on "minimal computer models" of LPS Re. The bulk density of monovalent and divalent ions in the simulation volume was consistent with the experimental buffer conditions (100 mM KCl and 0 mM or 50 mM $CaCl_2$). Details of the MC simulations are given elsewhere [8, 10, 16]. The ion density profiles of K^+ and Ca^{2+} predicted by the MC simulations for the absence and the presence of 50 mM $CaCl_2$ are

---

[5] Beyond these qualitative predictions the continuum description is not employed here. Especially it should not be used for the microscopic description, as it loses its validity for the high charge densities and ion strengths at play (see Sect. 4.2). Moreover, this description does not take the z-extension of the charged saccharides and the volume occupied by the headgroups into account.

Fig. 6.13 Ion density
distributions as they result
from the MC simulations
carried out by D.A. Pink and
B.E. Quinn. *Dashed line* K$^+$
ions in the absence of
calcium. *Solid lines* Ion
density distributions (K$^+$ and
Ca^{2+}) in the presence of
50 mM CaCl$_2$

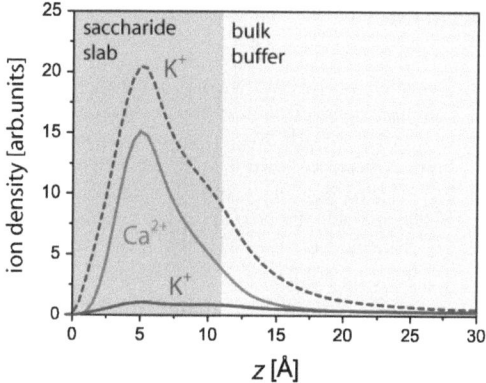

presented in Fig. 6.13. In the absence of calcium the K$^+$ concentration (dashed blue line) peaks right at the center of the saccharide slab at around 5 Å from the chain/saccharide interface, which is consistent with the experimental finding ($z_{max} < 10$ Å). In the presence of 50 mM CaCl$_2$, the Ca^{2+} concentration (solid red line) has a peak at the same position, while the monovalent K$^+$ ions (solid blue line) are almost completely displaced from the headgroup region. This is in excellent agreement with the presented experimental observations. Moreover, the MC simulations fully support the assumption that negative Cl$^-$ ions do not significantly contribute to the charge compensation, and that the above presented calculation of effective charge densities is valid.

6.3.5 Summary of Sect. 6.3

A model of the outer bacteria membrane was created by the preparation of a monolayer of purified rough mutant lipopolysaccharides (LPS Re) on buffers with and without calcium ions. The electronic structure of the monolayer was characterized using GIXOS. The vertical density profiles of monovalent (K$^+$) and divalent (Ca^{2+}) cations at the LPS monolayers in the absence and in the presence of 50 mM Ca^{2+} were quantitatively investigated using grazing incidence X-ray fluorescence. In the absence of calcium ions, the K$^+$ concentration peak was localized in the negatively charged LPS headgroup region. It was found that in sufficient amount (here 50 mM), divalent Ca^{2+} ions vastly replace the monovalent K$^+$ ions from the headgroup region. These observations were found to be in excellent agreement with the predictions of coarse-grained MC simulations of the ion density profiles near LPS Re monolayers, which were carried out by collaboration partners. Moreover, the vertical integration of the experimentally reconstructed excess ion density profiles provided the accurate measurement of the effective charge of the LPS Re molecules in the monolayer.

References

1. S. Snyder, D. Kim, T.J. McIntosh, Lipopolysaccharide bilayer structure: effect of chemotype, core mutations, divalent cations, and temperature. Biochemistry **38**, 10758 (1999)
2. U. Seydel, M.H.J. Koch, K. Brandenburg, Structural polymorphisms of rough mutant lipopolysaccharides Rd to Ra from Salmonella minnesota. J. Struct. Biol. **110**, 232 (1993)
3. E.A. Evans, Bending resistance and chemically induced moments in membrane bilayers. Biophys. J. **14**, 923 (1974)
4. J.N. Israelachvili, D.J. Mitchell, B.W. Ninham, Theory of self-assembly of hydrocarbon amphiphiles into micelles and bilayers. J. Chem. Soc. Faraday Trans. 2 **72**, 1525 (1976)
5. U. Seydel, M. Oikawa, K. Fukase, S. Kusumoto, K. Brandenburg, Intrinsic conformation of lipid A is responsible for agonistic and antagonistic activity. Eur. J. Biochem. **267**, 3032 (2000)
6. C.E. Miller, J. Majewski, T. Gog, T.L. Kuhl, Characterization of biological thin films at the solid-liquid interface by X-ray reflectivity. Phys. Rev. Lett. **94**, 238104 (2005)
7. E. Nováková, K. Giewekemeyer, T. Salditt, Structure of two-component lipid membranes on solid support: an x-ray reflectivity study. Phys. Rev. E **74**, 051911 (2006)
8. R.G. Oliveira et al., Physical mechanisms of bacterial survival revealed by combined grazing-incidence X-ray scattering and Monte Carlo simulation. Comptes Rendus Chimie **12**, 209 (2009)
9. D.A. Pink, L.T. Hansen, T.A. Gill, B.E. Quinn, M.H. Jericho, T.J. Beveridge, Divalent calcium ions inhibit the penetration of protamine through the polysaccharide brush of the outer membrane of Gram-negative bacteria. Langmuir **19**, 8852 (2003)
10. E. Schneck, E. Papp-Szabo, B.E. Quinn, O.V. Konovalov, T.J. Beveridge, D.A. Pink, M. Tanaka, Calcium ions induce collapse of charged O-side chains of lipopolysaccharides from pseudomonas aeruginosa. J. R. Soc. Interface **6**, S671 (2009)
11. V. Padmanabhan, J. Daillant, L. Belloni, Specific ion adsorption and short-range interactions at the air aqueous solution interface. Phys. Rev. Lett. **99**, 086105 (2007)
12. N.N. Novikova et al., X-ray fluorescence methods for investigations of lipid/protein membrane models. J. Synchrotron Rad. **12**, 511 (2005)
13. W. Bu, D. Vaknin, X-ray fluorescence spectroscopy from ions at charged vapor/water interfaces. J. Appl. Phys. **105**, 084911 (2009)
14. W.B. Yun, J.M. Bloch, X-ray near total external fluorescence method: experiment and analysis. J. Appl. Phys. **68**, 1421 (1990)
15. N.N. Novikova et al., Total reflection X-ray fluorescence study of Langmuir monolayers on water surface. J. Appl. Cryst. **36**, 727 (2003)
16. K. Kjaer, Some simple ideas on X-ray reflection and grazing-incidence diffraction from thin surfactant films. Phys. B **198**, 100 (1994)
17. K. Kjaer, J. Als-Nielsen, C.A. Helm, L.A. Laxhuber, H. Möhwald, Ordering in lipid monolayers studied by synchrotron X-ray-diffraction and fluorescence microscopy. Phys. Rev. Lett. **58**, 2224 (1987)
18. K. Kjaer, J. Als-Nielsen, C.A. Helm, P. Tippmankrayer, H. Mohwald, Synchrotron x-ray-diffraction and reflection studies of arachidic acid monolayers at the air-water-interface. J. Phys. Chem. **93**, 3200 (1989)
19. R.G. Oliveira et al., Crucial roles of charged saccharide moieties in survival of gram negative bacteria revealed by combination of grazing incidence x-ray structural characterizations and Monte Carlo simulations. Phys. Rev. E **81**, 041901 and successive pages (2010)
20. K. Brandenburg, U. Seydel, Physical aspects of structure and function of membranes made from lipopolysaccharides and free lipid A. Biochim. Biophys. Acta **775**, 225 (1984)
21. H. Labischinski, G. Barnickel, H. Bradaczek, D. Naumann, E.T. Rietschel, P. Giesbrecht, High state of order of isolated bacterial lipopolysaccharide and its possible contribution to the permeation barrier property of the outer membrane. J. Bacteriol. **162**, 9 (1985)

22. U. Seydel, K. Brandenburg, M.H.J. Koch, E.T. Rietschel, Supramolecular structure of lipopolysaccharide and free lipid A under physiological conditions as determined by synchrotron small-angle X-ray diffraction. Eur. J. Biochem. **186**, 325 (1989)
23. D.C. Grahame, Diffuse double layer theory for electrolytes of unsymmetrical valence types. J. Chem. Phys. **21**, 1054 (1953)
24. B.W. Ninham, V.A. Parsegian, Electrostatic potential between surfaces bearing ionizable groups in ionic equilibrium with physiologic saline solution. J. Theor. Biol. **31**, 405 (1971)
25. J.N. Israelachvili, *Intermolecular and Surface Forces* (Academic Press Inc., London, 1985)
26. R.T. Coughlin, A.A. Peterson, A. Haug, H.J. Pownall, E.J. McGroarty, A pH titration study on the ionic bridging within lipopolysaccharide aggregate. Biochim. Biophys. Acta **821**, 404 (1985)
27. S.O. Hagge, M.U. Hammer, A. Wiese, U. Seydel, T. Gutsmann, Calcium adsorption and displacement: characterization of lipid monolayers and their interaction with membrane-active peptides/proteins. BMC Biochem. **7**(1), 15 (2006)
28. J.N. Israelachvili, *Intermolecular and Surface Forces* (Academic Press Inc., London, 1991)
29. K. Hu, A.J. Bard, Use of atomic force microscopy for the study of surface acid-base properties of carboxylic acid-terminated self-assembled monolayers. Langmuir **13**(19), 5114–5119(1997)

Chapter 7
Conclusions

In this thesis, the structure of membrane-bound saccharides and their influence on the mechanical properties of membranes were studied using planar membrane models in combination with various X-ray and neutron scattering techniques.

In the first step, multilayer stacks of glycolipid membranes deposited on solid substrates were used as well-defined model systems to investigate the influence of membrane-bound saccharides on the mechanical properties of interacting membranes using specular and off-specular neutron scattering (Sect. 5.1). In order to quantitatively determine the compression and bending moduli of the interacting membranes from the experimental reciprocal space maps, a new modeling approach that accounts for the finite sample size was developed. A cut-off parameter that defines an integration limit for the calculation of membrane displacement correlation functions allows for comprehensive modeling of the scattering signals from membrane stacks, which is in contrast to previous approaches based on power-law decays or numerical back-transformations of integrated Bragg sheet intensities. The experimental results clearly indicated a significant influence of the saccharide headgroup structure on inter-membrane interactions and membrane mechanics. Furthermore, the glycolipid membranes exhibited a much weaker swelling than commonly studied phospholipid systems, suggesting an important role of attractive interactions mediated by the saccharide headgroups in fine-adjusting inter-membrane contacts.

In order to study specific saccharide–saccharide interactions important in cell recognition processes, glycolipids with LewisX groups were incorporated into phospholipid membrane stacks (Sect. 5.2). Specular and off-specular neutron scattering demonstrated that LewisX forms homophilic pairs that cross-link adjacent membranes and mechanically rigidify the membrane multilayers. A numerical calculation of the subtle balance of attractive and repulsive inter-membrane interactions (Sects. 5.2 and 4.2) yielded a lower limit of the forces and energies required to cross-link the membranes.

E. Schneck, *Generic and Specific Roles of Saccharides at Cell and Bacteria Surfaces,* Springer Theses, DOI: 10.1007/978-3-642-15450-8_7,
© Springer-Verlag Berlin Heidelberg 2011

Based on the results obtained with synthetic model systems, the strategy established above was further extended to more complex membranes of bacterial lipopolysaccharides. The mechanical properties of LPS membrane multilayers were for the first time determined by specular and off-specular neutron scattering (Sect. 6.1). The experiments showed that the bending rigidities of the LPS membranes (of the order of 1 $k_B T$) are much lower than those of commonly studied phospholipid membranes. Longer, more flexible saccharide headgroups resulted in a significantly softer inter-membrane confinement but at the same time with to a higher membrane bending rigidity. Moreover, the inter-membrane confinement and the stability of the multilayers were significantly enhanced by calcium ions.

To further investigate the influence of divalent cations on the conformation of LPS molecules, a realistic model of outer bacteria membranes was created by the deposition of monolayers of complex and polydisperse lipopolysaccharides extracted from *Pseudomonas aeruginosa* PAO1 onto a planar substrate functionalized with a hydrophobic alkyl silane monolayer. This model system was structurally characterized using specular X-ray reflectometry with high energy (22 keV) that provides for high transmittance through the bulk buffer solution. For the first time it was experimentally proven that Ca^{2+} ions induce a collapse of the negatively charged LPS *O*-sidechains towards the core saccharide region (Sect. 6.2). The conformational change is in excellent agreement with the predictions of coarse-grained Monte Carlo simulations that postulate a key role of divalent cations in the defense mechanism of Gram-negative bacteria against antimicrobial peptides.

In order to determine the amount and location of divalent cations at bacterial surfaces, grazing-incidence X-ray fluorescence (GIXF) was applied to mutant LPS monolayers deposited at the air/water interface (Sect. 6.3). For this purpose, a method for the interpretation of X-ray fluorescence intensities from biological model systems at the air/water interface was developed (Sect. 4.3). An algorithm for the calculation of X-ray illumination profiles from a slab model representation of the studied systems was theoretically derived and the implications of X-ray absorption were systematically investigated. In the absence of divalent Ca^{2+} ions, a concentration peak of the monovalent K^+ ions was found in the negatively charged LPS headgroup region. In the presence of divalent Ca^{2+} ions, monovalent K^+ ions are displaced from this region. Moreover, the experiments provided the accurate measurement of the average number of monovalent and divalent cations associated with one LPS Re molecule.

The results presented in this thesis demonstrate that specular and off-specular X-ray and neutron scattering from well-defined planar model systems provides quantitative insight into the structure and mechanics of complex biological surfaces.

Chapter 8
Outlook

The presented combination of various X-ray and neutron scattering techniques on the one hand together with the use of oriented model systems on the other hand holds much promise for the systematic study of complex biological surfaces.

From a fundamental viewpoint, the shown strategy can be readily extended in order to study structure, mechanics, and dynamics of a broad variety of biological surfaces. This will allow for the systematic investigation of biologically highly relevant processes taking place at cell and bacteria surfaces, such as the mode of action of membrane-active drugs.

From the application side, the presented approach will for instance enable the comprehensive out-of-plane and in-plane characterization of wet electrochemical biosensors, where solid-supported membranes allow integration of biological soft matter functionality with hard semiconductor or metal devices.

Generally, in order to fully explore the potential of scattering experiments, modeling the measured scattering signals in DWBA can be used to extract structural information from the entire range of the recorded reciprocal space maps. This will be particularly important for the study of samples with weaker scattering length density contrast and/or weaker structural ordering (such as planar, supported membranes interacting with soft polymer interlayers), where relevant scattering features may be only located in the regions that cannot be treated in a kinematic approximation. Such scattering signals will be even more valuable if they are integrated with real-space computer simulations of the studied biological surfaces. Continuum-mechanical models as well as coarse-grained Monte Carlo simulations and atomistic molecular dynamics simulations provide a valuable estimate for the model parameterization of the scattering signals. Here, especially multi-scale approaches, where simulations representing molecular details are utilized to establish a coarse-grained or continuum-mechanical description of the systems, will have a great potential. Conversely, the computer models themselves can be continuously improved through detailed comparison with the experimental results.

E. Schneck, *Generic and Specific Roles of Saccharides at Cell and Bacteria Surfaces,* Springer Theses, DOI: 10.1007/978-3-642-15450-8_8,
© Springer-Verlag Berlin Heidelberg 2011